Foreword

The information contained in this book has been compiled through personal experience over

many years. The book is designed to be an important part of the toolbox as a source of reference

both during and after the apprenticeship stage. Although it would be impossible to give an

example for every problem that may be encountered throughout every boilermaker's career, the

principles laid out here should provide enough of a guide line to allow for the manipulation of

every process described, and should enable the boilermaker to conquer the vast majority. The

many formulae used in this book have not been discovered or invented by myself, but rather

handed down through countless generations by lofty mathematicians from ancient times.

My advice to those who encounter complicated drawings with the necessary sizes or angles

seemingly impossible to find is;

"Slow down, have a little patience and look carefully for the triangle. I promise you it will be

there!" Always remember: "Check twice, cut once".

The Author.... Jim Draper.

The Mathematics of Boilermaking

by

Jim Draper

CHAPTER 4. Basic Geometry and Trigonometry.

CHAPTER 5. Pyramids and Cones.

CHAPTER 6. Transitions.

CHAPTER 1

Common geometric shapes and their properties

The first chapter is an introduction to the geometric shapes and the laws governing them that the

boilermaker will, throughout his career, become both

familiar and comfortable with.

1a. Angles.

A line perpendicular to another line produces two equal

angles of **90°** and are called right-angles,

Fig 1-1a. The line **OC** is perpendicular to **AB** and the

angles at **O** either side of the perpendicular are both **90°**

which is always shown as a square in the corner.

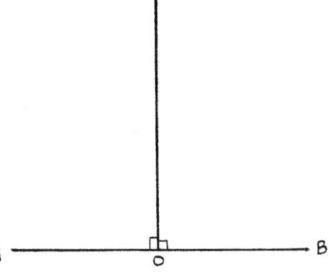

Fig 1-1a Perpendicular

An angle less than 90° is an acute angle, **Fig 1-2a.**

The angle between lines **OB** and **OC** is acute.

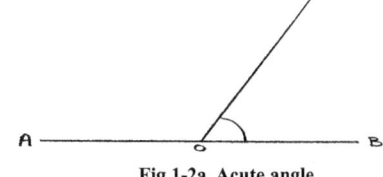

Fig 1-2a. Acute angle

An angle greater than 90° is an obtuse angle, **Fig 1-3a.** The angle between the lines **OA** and **OC**

is obtuse.

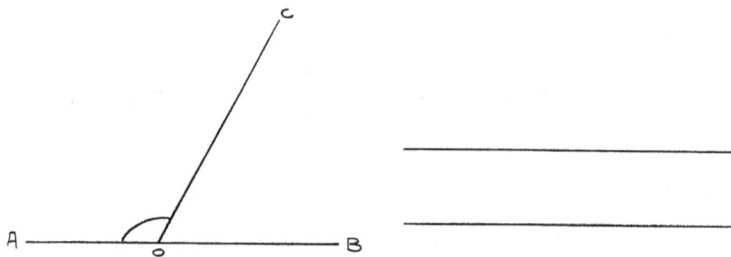

Fig 1-3a Obtuse angle　　　　　　**Fig 1-4a Parallel lines**

Two lines that remain equidistant from each other can never produce an angle and are called

parallel lines, **Fig 1-4a.**

1b. Rectangle.

A rectangle is a figure with four sides and

four right-angles. The opposite sides are

equal.

 Fig 1-1b. If all four sides are equal it is said

to be a square. The sum of the four angles is

360°.

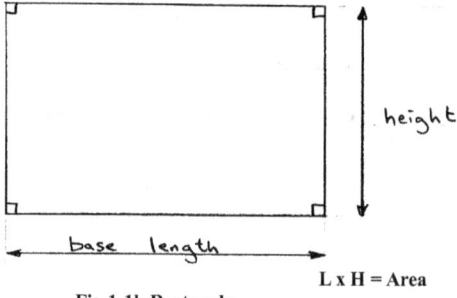

Fig 1-1b Rectangle

L x H = Area

The area of a rectangle can be determined by multiplying its base length by its height.

AREA will always be a square of the unit used: square inches; square feet; square

millimetres; square meters; etc.

Example. A rectangle has a base length of 60 and a height of 35.

Area = length x height

Area = 60 x 35

Area = 2100 square units.

1c Triangle.

A triangle is a figure with three sides and three angles. **Fig 1-1c.** The sum of the three angles is **180°**. The area of a triangle can be determined by multiplying its base length by its height and then dividing the answer by 2.

$$Area = \frac{L \times H}{2}$$

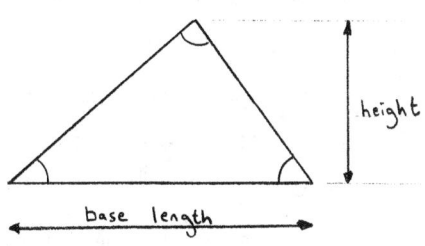

Fig 1-1c Triangle

Example. A triangle has a base length of 90 and a height of 42.

$$Area = \frac{length \times height}{2}$$

$$Area = \frac{90 \times 42}{2}$$

$$Area = \frac{3780}{2} \quad \text{...........} \quad Area = 1890 \text{ square units.}$$

1d Circle.

A circle is a line without a beginning or an end called a **circumference** on which all points are equidistant from a centre point. A circle encompasses **360°, Fig 1-1d.**

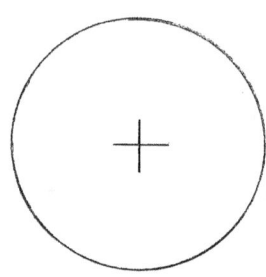

Fig 1-1d Circle

A line passing through the centre point and touching the circumference at two points is the **diameter.**

A line from the centre point to a point on the circumference is the **radius**, which is equal to half the diameter. In **Fig 1-2d, AOB** is the diameter and **OB** or **OA** is the radius.

The area of a circle can be determined by multiplying its radius squared, which means multiplied by itself, by **π**.

π is the symbol used for the ratio of the circumference of a circle to its diameter.

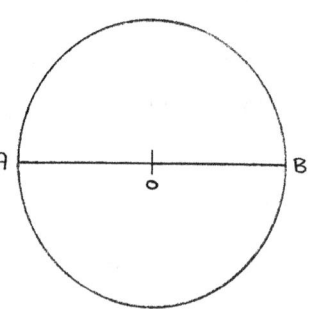

Fig 1-2d Diameter and radius

The numerical value of **π** is **3.141592654...............** The first 5 numbers after the decimal point are sufficient in most calculations.

Example. A circle has a diameter of 84 therefore its radius is 42.

$$\textbf{Area} = \boldsymbol{\pi r^2}$$

Area = 3.14159×42^2

Area = 3.14159×1764..... Area = 5541 square units.

The circumference of a circle can be determined by multiplying its

diameter by **π**.

In **Fig 1-3d** the diameter **AB** is multiplied by **π** to find the

circumference.

Example. A circle has a diameter of 84

Circumference = πD ... Circumference = 3.14159 x 84

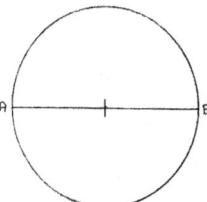

Fig 1-3d Circumference

$$\text{Circumference} = 263.89 \text{ units.}$$

A line outside a circle touching the circumference at one point

is a **tangent**.

A line inside a circle touching the circumference at two points

is a **chord. Fig 1-4d**.

AB is a chord.

CD is a tangent.

The shaded area in the figure is a segment of the circle.

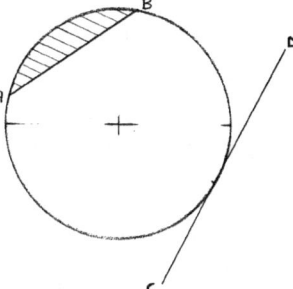

Fig 1-4d Chord and tangent

The radius of a circle produces six equal chords on its circumference **Fig 1-5d**, therefore **O** to any point, such as **OA** or **OC;** etc is equal to **AB** or **CD,** etc. Since a circle is made up of **360°** each of the six triangles inside the circle has an apex angle of **60°**...... $\frac{360}{6}$ = 60.

If all sides are equal in length then all angles must also be equal, i.e. **60°**.

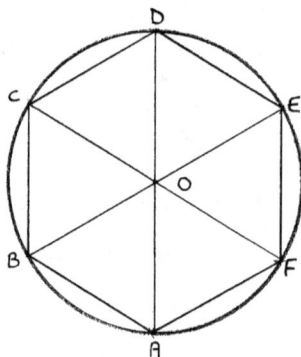

Fig 1-5d Radius chords

A section of a circle between two radius lines is a **sector,** and the section of the circumference enclosing the sector is an **arc. Fig 1-6d.**

The shaded area **AOB** is a sector.

Circumference part **A** to **B** is an arc.

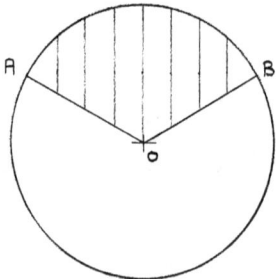

Fig 1-6d Sector and arc

1e Polygon.

A polygon is a figure having more than four sides. When all sides are equal it is a regular polygon.

The area of a polygon can be determined by adding together the areas of all its triangles.

Fig 1-1e.

To determine the area of a regular polygon, it is necessary to determine the area of one triangle only since they are all identical, and multiply the answer by the number of triangles.

Area of triangle x 5 = area of pentagon.

$$area = \frac{L\,x\,H}{2} \times 5$$

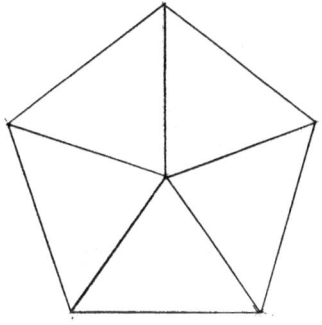

Example. If the area of triangle of a regular pentagon (5 sides) is known to be 1400 square units.

Area = 1400 x 5   ....

Area of pentagon = 7000 square units.

Fig 1-1e Polygon

CHAPTER 2

Marking out

Marking out will occupy a great deal of the boilermaker's time during his career and over time he will become quite a proficient draughtsman. As in all things the absolute basics must first be mastered.

2a 90°.

Example.

To mark out a perpendicular line to any line; Draw a horizontal line, then, with a pair of dividers set to any size, mark an arc to produce points **A** and **B** from point **O**. **Fig 2-1a.**

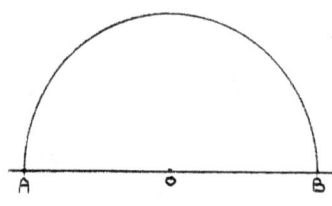

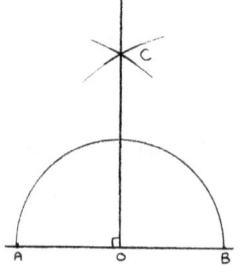

Fig 2-1a Perpendicular Step 1

Fig 2-2a Perpendicular Step 2

Open the dividers wider and mark two intersecting arcs above point **O** using **A** and **B** as centre points. The point of intersection is point **C**. A line drawn from point **O** through point **C** will be at right-angles to the base line, **Fig 2-2a.**

2b 45°.

Example.

To mark out a line at 45° to any line, mark a

perpendicular to that line as in the previous exercise.

Fig 2-1b. Using points **D** and **B** as centres mark two

intersecting arcs crossing at point **E**. A line drawn

from point **O** through point **E** will be at **45°** to the

base line.

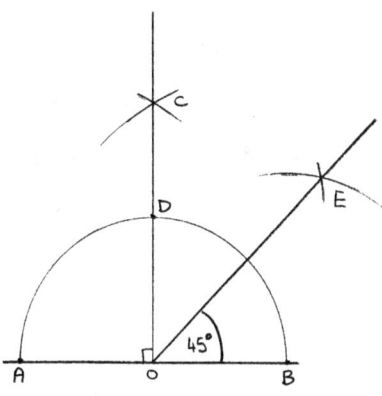

Fig 2-1b Marking 45°

2c Degrees of a circle.

Example for 30°.

To mark degrees of a circle first describe a semi-

circle on a base line, **Fig 2-1c**.

Mark a perpendicular from centre point **O** as before

to produce point **C**.

With the dividers set at radius size **OA** use points **B**

and **C** as centres to describe intersecting arcs at

points **D** and **E**. Radial lines drawn through these

points from centre point **O** will be in graduations of

30°.

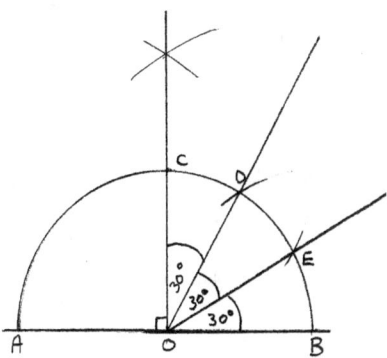

Fig 2-1c Marking 30°

Example for 15°.

In **fig 2-1c,** if two points adjacent to each other

such as **B** and **E** are bisected an angle of **15°**

will be produced as in **Fig 2-2c.**

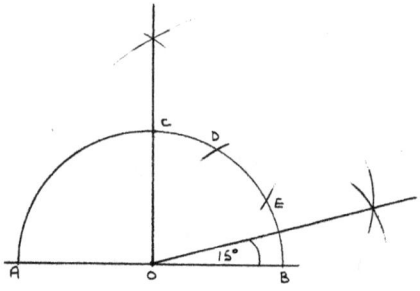

Fig 2-2c Marking 15°

Any chord can be divided into equal parts by means of **parallel lines**.

Example.

In this example the chord **A B** is divided

into four equal parts **fig 2-3c.**

From point **A** step off four equal arcs

with the dividers. With the dividers set at

the distance from point **A** to point **4** draw

an arc from point **B.**

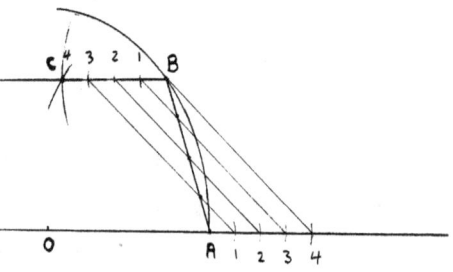

Fig 2-3c Divide a chord

With the dividers set at the distance from point **B** to point **4** draw an arc from point **A** to intersect

at point **C.**

Draw in the line **CB,** which will be parallel to the base line and divide it into the same 4 divisions

as used in **A** to **4.**

Draw the parallel lines connecting the opposite points; **B** to **4, 1** to **3,** etc.

The intersection points on the chord **AB** will produce four equal spaces on that chord.

If radial lines from point **O** are drawn through these intersection points, the graduations will have

degrees of the same value, **providing the chord has an arc value of 30 ° or less.**

Example for 5°.

An angle of **15°** can

be divided into three

equal parts of **5°** each

using parallel lines,

Fig 2-4c. From point

B with the dividers

set at any size, step

off three equal parts

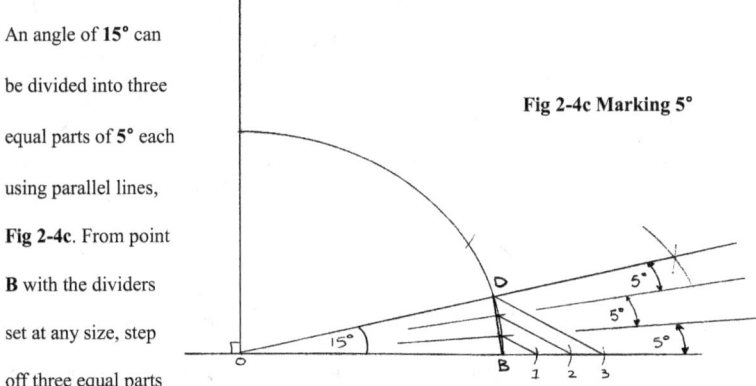

Fig 2-4c Marking 5°

numbered 1 to 3. Draw a line connecting point **3** to point **D**, and draw the next two lines parallel

to this one to give the intersection points on the chord **BD**. Radial lines drawn from point **O**

through these intersection points will be in graduations of **5°**.

Example for 1°.

An angle of **5°** can be divided into five equal parts of 1° each using parallel lines as before,

Fig 2-5c.

From point **B** with the dividers set at any size, step off five equal parts numbered 1 to 5.

Draw a line connecting point **5** to point **C**, and draw the next four lines parallel to this one to give

the intersection points on the chord **BC**. Radial lines drawn through these points from the centre point will be in graduations of **1°**.

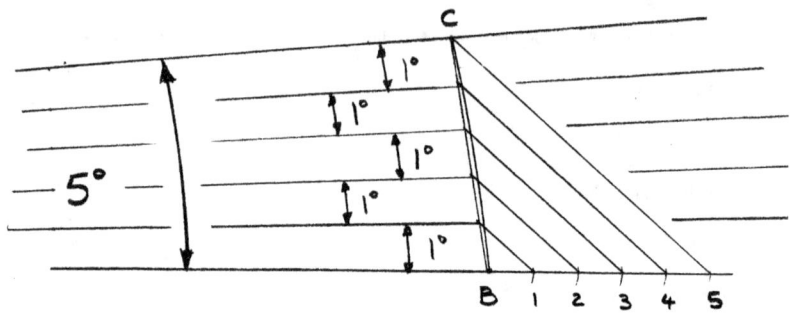

Fig 2-5c Marking 1°

Any angle between 0° and 90° can be quickly marked out using the trigonometry shown in

chapter 4, the application of the tangent formula being the most suitable, as the base line can be

any length with the opposite side giving the required height to mark the desired angle.

2d Rectangle.

Marking out a rectangle requires the use of parallel lines.

Example.

Fig 2-1d shows a rectangle 700 x 500.

Step 1 is to mark two horizontal parallel lines 500 apart.

Step 2 is to mark a perpendicular joining the two parallel lines.

Step 3 is to measure and mark a point on each parallel line at a distance of 700 from the perpendicular, and draw in the fourth line between these two points.

Opposite sides should be equal and parallel and all angles should be right-angles. This can be checked by measuring across the points from both opposite corners which should both have the same measurement, and can be found by using the trigonometry in chapter 4 to be a measurement of 860.2.

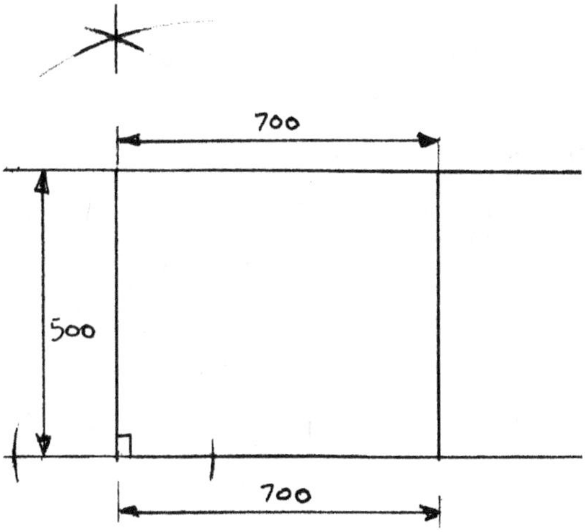

Fig 2-1d Marking out a rectangle

2e Triangle.

The side opposite angle A will be side a.

The side opposite angle B will be side b.

The side opposite angle C will be side c.

If a triangle is to be marked out and the length of all sides is known, it is done using **intersecting arcs.**

Example.

Fig 2-1e shows a triangle of sides; c = 900,

b = 800, a = 500.

Step 1 is to draw the line **AB** with the

distance from **A** to **B** being 900.

Step 2 is to set the dividers to a value of 800

and describe an arc using point **A** as a centre.

Step 3 is to set the dividers to a value of 500

and describe an arc using point **B** as a centre

to give the intersection point **C**.

Step 4 is to draw in the lines **AC** and **BC**.

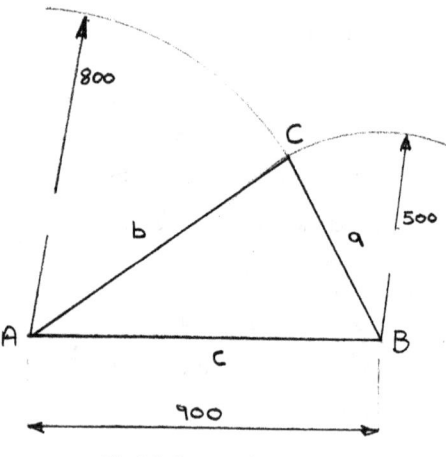

Fig 2-1e Intersecting arcs

If a triangle is to be marked out with the length of one side

known and the two angles common to that side known it is

done using **intersecting lines**.

Example.

Fig 2-2e shows a triangle with side c = 600, angle A = 47°,

angle B = 63°.

Step 1 is to draw the line **c** with a distance from **A** to **B**

being 600.

Step 2 is to draw a line from point **A** with the use of a

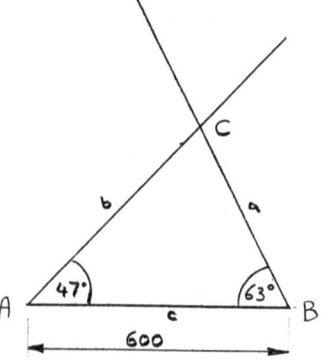

Fig 2-2e Intersecting lines

protractor at an angle of 47° to line **c**.

Step 3 is to draw a line from point **B** with the use of a protractor at an angle of 63° to line **c**.

The two lines will intersect at point **C** to complete the triangle.

Obviously; triangles can also be drawn using a combination of intersecting lines and intersecting arcs.

CHAPTER 3

Material lengths

3a. Straight length with bends.

When marking out plates or flat bar for bending where sharp bends are required, only inside sizes are used, **od** stands for outside dimension and **id** stands for inside dimension.

Example 1.

Fig 3-1a a bracket is 740 od long x 500 od high x 65 wide x 12 thick.

The plate needed to make the bracket will be:

(740 - 12) = 728 + (500 - 12) = 488.

728 + 488 = 1216. ....

1216 x 65 x 12, plate size before bending.

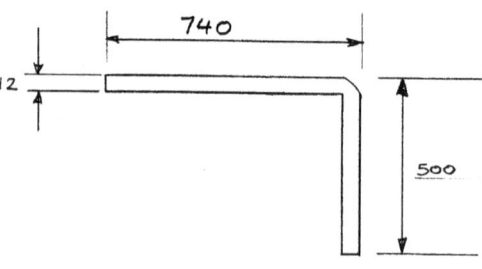

Fig 3-1a Straight length with one bend

Example 2.

Fig 3-2a has three bends but still follows the rule of only using inside sizes. The example can be dealt with as four separate sections; parts A, B, C and D.

The material used is 65 wide and 20 thick. Careful note must always be taken as to whether the dimensions given are inside or outside sizes or a combination of both.

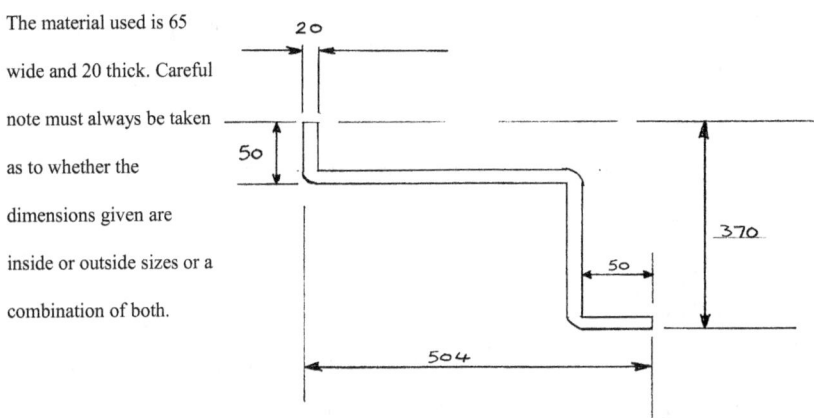

Fig 3-2a Straight length with multiple bends

In **fig 3-3a** the material needed for part A is .... 50 - 20 = 30

The material needed for part B is.. 504 - 50 - 20 - 20 = 414

The material needed for part C is .... 370 - 50 - 20 = 300

The material needed for part D is 50

The length of material needed for the example is

.... 30 + 414 + 300 + 50 = 794

The plate size of the material before bending is

.... 794 x 65 x 20.

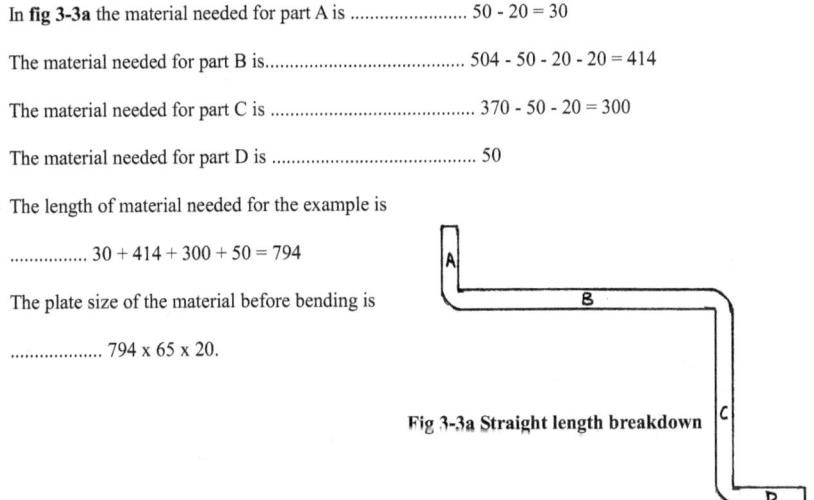

Fig 3-3a Straight length breakdown

It should be noted that two bends are marked on one side of the material and one is marked on the opposite side to make this bracket.

If this same bracket was required to have more strength at the bends, it would have been constructed with **radius bends equal to the material thickness,** which would then require both straight and circular length to be calculated as dealt with in part **3c.**

3b. Circular length using the mean.

The **'mean'** is the name given to the average diameter of a rolled plate or pipe. It is determined by either adding the material thickness to the inside diameter or subtracting the material thickness from the outside diameter, which of course will produce the same answer.

Example 1.

Fig 3-1b. A plate 100 wide and 20 thick is rolled to an inside diameter of 500.

When rolling plates or flat bar, the **mean diameter** of the finished article must be multiplied by **π** to give the required cutting length.

In this example the mean is 500 + 20 = 520..............

520 x π = 1633.62.

The material needed for this example is ....

1633.6 x 100 x 20.

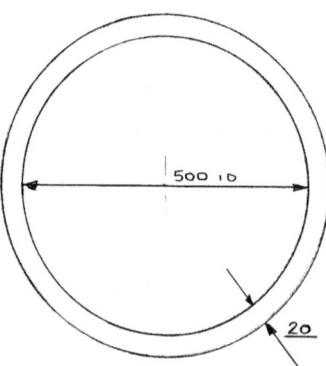

Fig 3-1b Circular length

Example 2.

A shell plate for a vessel (vessels usually consist of many plates welded together known as shells)

is 3000 wide and 28 thick and 5400 id.

Mean = 5400 + 28 = 5428

Cutting length = π x 5428........... 3.14159 x 5428 = 17052.55. This can be rounded off to 17052.

At this point it should be mentioned that it is good practice to mark the four cardinal

points onto the plate before rolling.

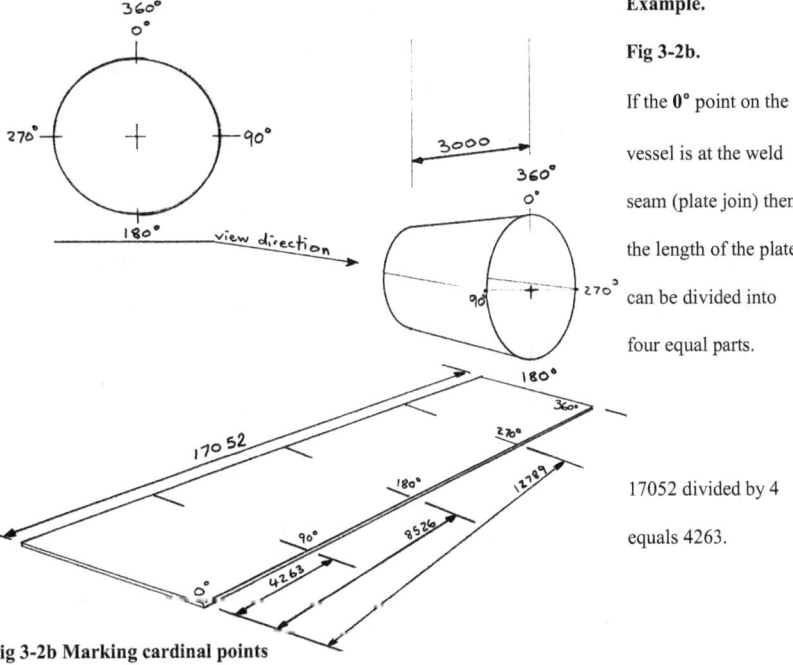

Example.

Fig 3-2b.

If the **0°** point on the vessel is at the weld seam (plate join) then the length of the plate can be divided into four equal parts.

17052 divided by 4 equals 4263.

Fig 3-2b Marking cardinal points

If the end of the plate is **0°**, then **90°** will be at 4263 from the end of the plate.

180° will be 4263 further on at 8526

270° will be 4263 further on at 12789

360° will be 4263 further on at 17052 which is the other end of the plate. **0°** and **360°** are the same point when the plate is rolled.

3c Combinations of straight and circular.

The lessons learned in 3a and 3b still apply when using them in conjunction with each other. Therefore only inside sizes are used when marking the position of a sharp bend and the mean is used for the rolled or radius sections.

Example 1.

Fig 3-1c a bracket 50 wide and 20 thick has 4 radius bends equal to the material thickness. It has an inside measurement of **200** and an outside measurement of **340**, with its height being **160.**

The material thickness is **20**, therefore the inside radius of the bends is also **20.**

Radius 20 multiplied by 2 equals 40 diameter.

40 diameter plus 20 thickness equals **60 mean diameter.**

There are 4 bends of 90° each with a combined sum of 360°.

60 x π = 188.49. .... $\frac{188.49}{4}$ = **47.122** which can be rounded off to **47.**

47 is the material length needed for one radius bend.

The combined length of the radius sections is 188.

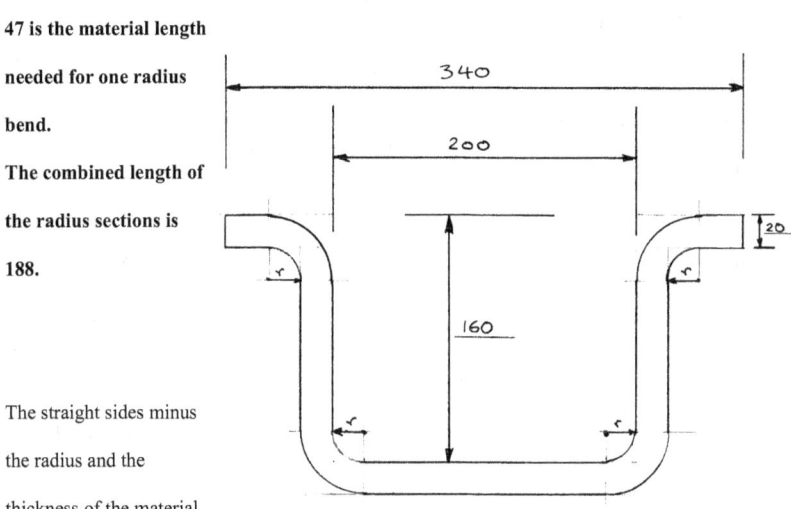

Fig 3-1c Radius bends

The straight sides minus the radius and the thickness of the material between each radius can now be added together as:

One centre section of 200 minus a radius on each end..... 200 - 20 -20 = 160,

Two long legs each minus a radius on each end and one thickness ...... 160 - 20 -20 -20 = 100,

Two short legs of.... 340 - 200 = 140 ...... divided by 2 = 70.........

70 minus one radius and one thickness ...... 70 -20 -20 = 30.

Total length of straight sides = 160 + 100 + 100 + 30 + 30 = 420.

Total length of material required is ...... 188 + 420 = 608.

Example 2.

Fig 3-2c a plate 100 wide and 24 thick is rolled to produce a 90 degree angle with an inside radius of 320 and a straight section of 140 on each leg.

Solution............ The inside

radius is 320 which when

multiplied by 2 gives an inside

diameter of 640.

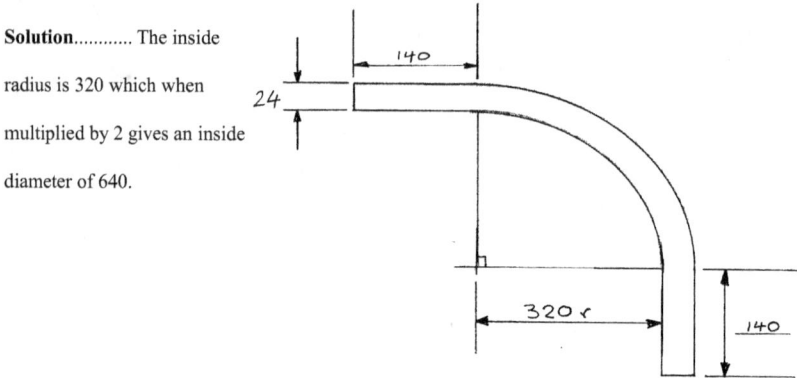

Fig 3-2c Circular and straight length

The **mean** is 640 + 24 = 664.................................... 664 x π = 2086.

2086 is the length of material needed to roll a full circle, but since the example calls for 90

degrees which is one quarter of a circle, the value 2086 is divided by 4 to give 521.5 which is the

material needed for the rolled section. The straight sections are both 140 so the total length of

material needed to produce the item is ....140 + 521.5 + 140 = 801.5.

Therefore the material needed is ...... 801.5 x 100 x 24.

Example 3.

Fig 3-3c. A double bracket is required to hold two pipes next to each other.

This is more complicated but the same rules still apply, the example must be taken step by step through the rolled sections and the straight sections as in the previous example. The final stage is simply to add all the sections together to find the total length of material required. (note the sharp bends)

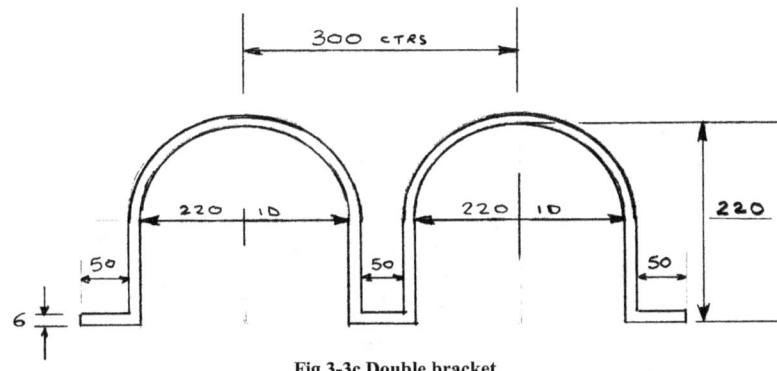

Fig 3-3c Double bracket

The rolled sections are both 220 inside diameter and encompass 180° each.

Mean = 220 + 6 = 226 .... 226 x π = 709.99999.... which we can call 710

This is the material needed for a full circle, which we divide by 2 for 180° to give 355.

The four vertical legs are all the same size.

The height from the bottom of the bracket to the inside of the radius is 220

The radius is 110 which we subtract from 220 to give 110, from which we must subtract the material thickness of 6 to give a final answer of 104.

The three base feet all have an inside size of 50.

Therefore the total length of 50 x 6 flat bar needed to make the bracket will be,..........

2 rolled sections 355 long, each = 710

4 vertical legs 104 long, each = 416

3 horizontal feet 50 long, each = 150

710 + 416 + 150 = **1276.**

To mark this flat bar for rolling and bending both sides need to be marked as in **fig 3-4c.**

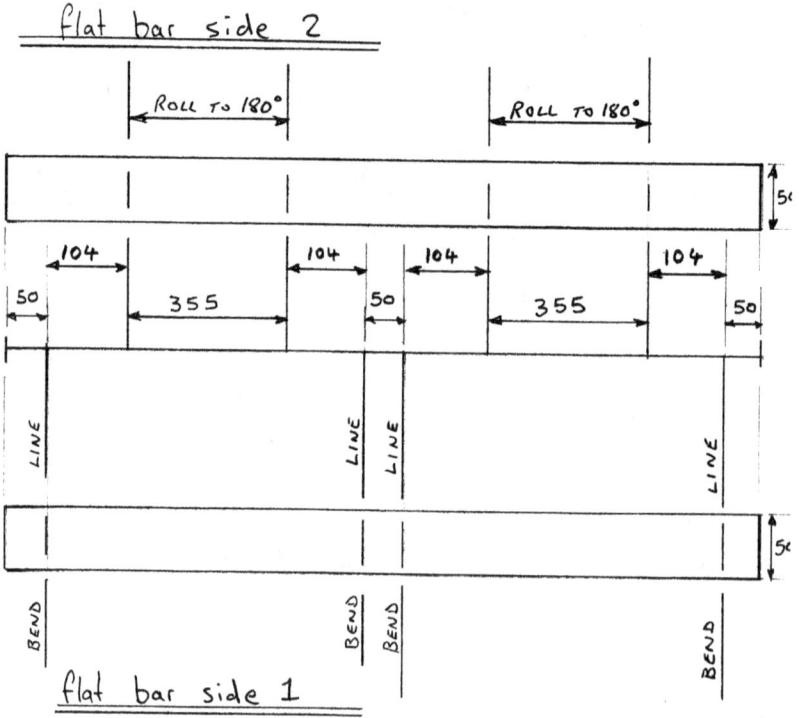

Fig 3-4c Bending and rolling lines

CHAPTER 4

Basic geometry and trigonometry

4a Basic Geometry.

It is important to be able to recognise the different triangles and apply their properties when interpreting drawings for marking out a layout, or a plate for cutting, bending or rolling.

A sound knowledge of basic geometry and trigonometry is as important to the boilermaker as the hammer or a piece of chalk in his tool box.

Many of the sizes needed to fabricate certain jobs will need to be calculated from the information provided in the drawings he will be working from:

Since a triangle is made up of three angles with a combined sum of **180°**, if any two angles are known;

the third angle can be found by subtracting the two known angles from **180°**, as is shown in **Fig 4-1a.**

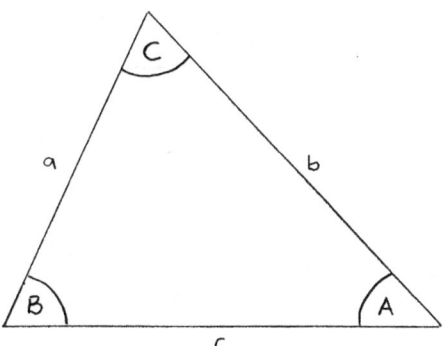

Fig 4-1a Any triangle

A = 36°...... B = 70°.

C = 180 - (36 + 70) = 74°.

If all three sides of a triangle are equal in length it is said to

be an **equilateral triangle**, and if all sides are equal it

follows that all angles are equal; namely **60°, Fig 4-2a.**

A = 60°

B = 60°

C = 60°

a = b = c.

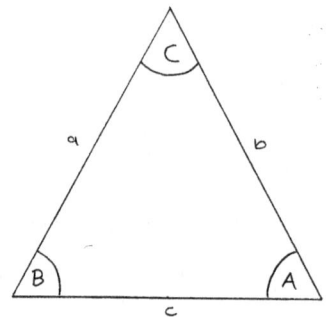

Fig 4-2a Equilateral triangle

If a triangle has two sides of equal length it is said to be an **isosceles triangle**, and if two sides

are equal it follows that the two angles opposite those sides are also equal,

Fig 4-3a.   A = B. a = b.

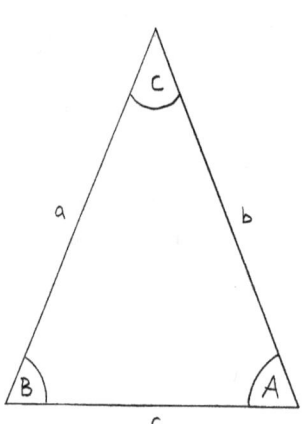

Fig 4-3a Isosceles triangle

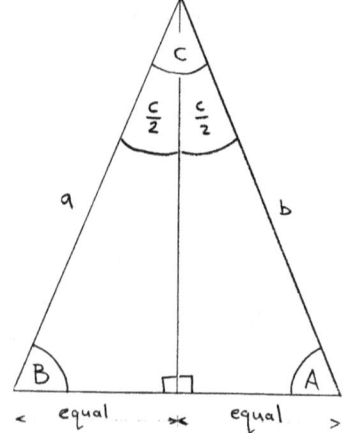

Fig 4-4a Isosceles triangle with perpendicular

In an isosceles triangle, the two equal angles are opposite the two equal sides. Therefore if a perpendicular line is drawn from the mid-point of the third side it will bisect the third angle, creating two identical right-angled triangles, with the perpendicular being common to both triangles.

Fig 4-4a. The triangle on the left is identical to the triangle on the right because; **A = B. a = b.**

Fig 4-5a Right-angled triangle

If one angle of a triangle is **90°** it is said to be a **right-angled triangle** and is subject to the **Pythagorean theorem; $a^2+b^2=c^2$**. The triangle will have a short side, a longer side and a longest side known as the **hypotenuse. Fig 4-5a.** Pythagoras states: The length of the short side squared, or multiplied by itself, added to the length of the longer side squared or multiplied by itself, is equal to the length of the hypotenuse squared or multiplied by itself.

This is easy to see if the triangle is drawn with the squares in place. **Fig 4-6a.**

This example shows $3^2 + 4^2 = 5^2$. Therefore $5^2 = 25$ and $\sqrt{25} = 5$

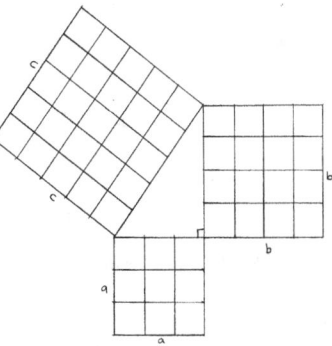

Fig 4-6a Pythagorean triangle

If all three sides of a triangle are unequal in length it is said to be a **scalene triangle**, and if all sides are unequal it follows that all angles are unequal.

Fig 4-7a.

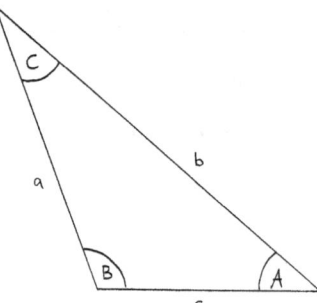

Fig 4-7a Scalene triangle

If two straight lines intersect, the opposite angles formed are **equal. Fig 4-8a.**

A = C. B = D.

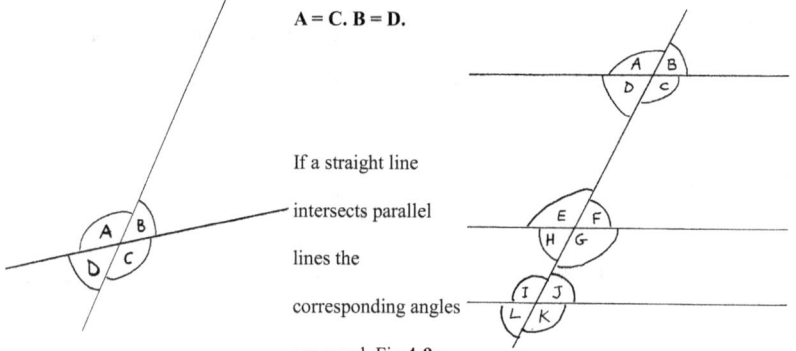

Fig 4-8a Intersecting lines

If a straight line intersects parallel lines the corresponding angles are equal. Fig **4-9a.**

A = C = E = G = I = K.

B = D = F = H = J = L.

Fig 4-9a Intersection of parallel lines

4b Basic Trigonometry.

Solving the right-angled triangle: In all formulae used; **Sin = Sine, Cos = Cosine,**

Tan = Tangent.

opp = opposite side, adj = adjacent side, hyp = hypotenuse. The formulae to be used are:

$$\sin = \frac{opp}{hyp} \qquad \cos = \frac{adj}{hyp} \qquad \tan = \frac{opp}{adj}$$

These formulae can be manipulated by simple substitution of known factors, such as; $3 = \frac{6}{2}$. In

this case involving the sine formula; sin has been replaced by 3, opp has been replaced by 6 and

hyp has been replaced by 2. Therefore

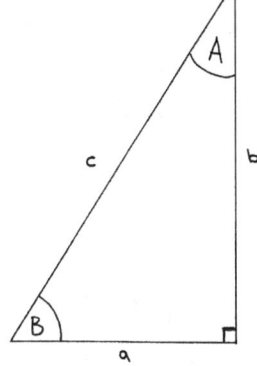

opp = sin x hyp = 6 = 3 x 2 and $hyp = \frac{opp}{\sin} = 2 = \frac{6}{3}$. All sides

will be referred to as: a, b or c, and their opposite angles will be

referred to as: A, B or C.

Adjacent refers to the side next to an angle excluding the

hypotenuse. **Fig 4-1b. b is the adjacent side to A and**

a is the adjacent side to B.

Opposite refers to the side opposite an angle. Fig 4-1b.

a is the opposite side to A and b is the opposite side to B.

Fig 4-1b Sides of the right-angle triangle

Hypotenuse refers to the side opposite the **90** degree angle and is always the longest side.

Fig 4-1b. Side c is the hypotenuse.

Example.... Fig 4-2b.

Side **b = 250**   angle **B = 28°**

Fig 4-2b Example using sine

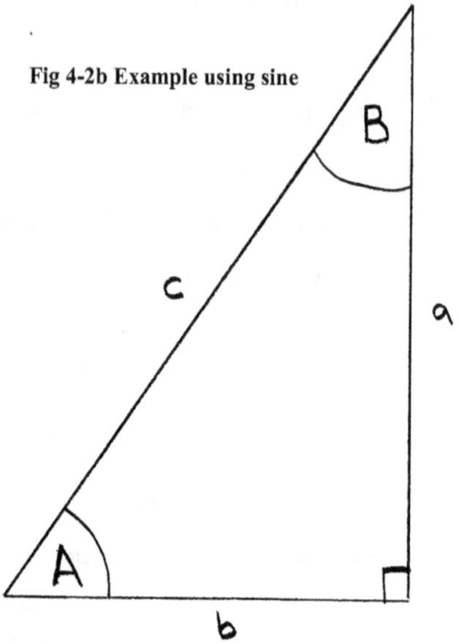

Since the **opposite** side of the known angle is

given we can use $\sin = \dfrac{opp}{hyp}$ and convert it to read

$hyp = \dfrac{opp}{\sin}$ and insert the known values into the

formula .... $hyp = \dfrac{250}{\sin 28}$ to give the answer

532.51

The length of the hypotenuse squared minus the

length of the second side squared will produce the length of the third side squared........

532.51² - 250² = 221066.9

The square root of this answer will give the length of the third side .... $\sqrt{221066.9}$ = **470.177**

side **c = 532.51**

side **b = 250**

side **a = 470.177**

angle **B = 28°**

angle **A = 90 - 28 = 62°**

Solving all other triangles: In triangles other than right-angled triangles, **the law of cosines,** or as it is sometimes called, **the cosine rule,** is used: $a^2=b^2+c^2-2bc$ **Cos A.**

The formulae for sides b and c are............ $b^2=a^2+c^2-2ac$ **Cos B.**

$$c^2=a^2+b^2-2ab \text{ Cos C.}$$

These formulae are used to determine length of side. To determine angle value the formulae are constructed thus...

$$\cos A = \frac{b^2 + c^2 - a^2}{2bc}$$

$$\cos B = \frac{a^2 + c^2 - b^2}{2ac}$$

$$\cos C = \frac{a^2 + b^2 - c^2}{2ab}$$

Together with **the sine rule;** .... $\dfrac{a}{b} = \dfrac{\sin A}{\sin B}$

and the knowledge that every triangle is made up of **180°,**

all triangles can be solved.

Since it can be daunting to apply the formula in its correct form to the many solutions required, examples have been provided with each relevant formula:

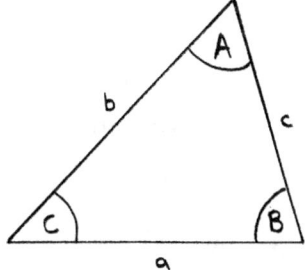

One side and two angles known. **Fig 4-3b.**

Side a is known, angle A is known, angle B is known.

Fig 4-3b One side and two angles known

$C = 180° - (A + B)$

$$b = \frac{a \times \sin B}{\sin A}$$

$$c = \frac{a \times \sin C}{\sin A}$$

Two sides and the common angle known. **Fig 4-4b.**

Side a is known, side B is known and angle C is known.

B = 180° - (A + C)

$$\tan A = \frac{a \times sinC}{b - (a \times cosC)}$$

c = √a²+b²- (2ab x cos C)

c can also be found by $c = \dfrac{a \times \sin C}{\sin A}$

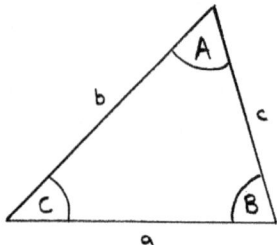

Fig 4-4b Two sides and common angle known

Two sides and one opposite angle known. **Fig 4-5b.**

Side a is known, side b is known and angle A is known.

C = 180° - (A + B)

$$\sin B = \frac{b \times \sin A}{a}$$

$$c = \frac{a \times \sin C}{\sin A}$$

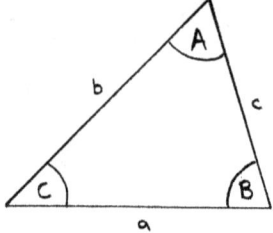

Fig 4-5b Two sides and one opposite angle known

Three sides known. **Fig 4-6b.**

Side a is known, side b is known and side c is known.

$$\cos A = \frac{b^2 + c^2 - a^2}{2bc}$$

$$\sin B = \frac{b \times \sin A}{a}$$

$$C = 180° - (A + B)$$

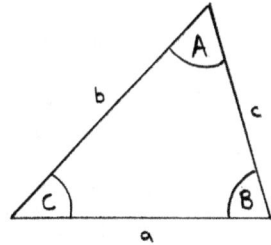

Fig 4-6b Three sides known

A working example is shown in **Fig 4-7b.** In this example one side and two angles are known:

a = 80, A = 65° and B = 75°.

With this information the value of the other two sides and the third angle can be found.

C = 180° - (65 + 75) = 40°

$$b = \frac{80 \times \sin 75}{\sin 65} \text{ .... } b = 85.2624$$

$$c = \frac{80 \times \sin 40}{\sin 65} \text{ .... } c = 56.739$$

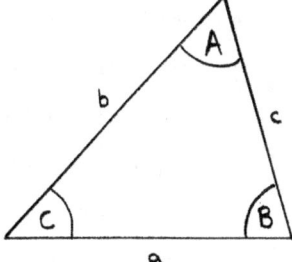

Fig 4-7b Example of one side and two angles known

Knowing all sides and all angles in the previous

example it is now a simple matter to determine the

height of the triangle by inserting a perpendicular on

line a to meet the apex of **angle A. Fig 4-8b.**

The perpendicular is **line xy** and it will be seen that

it forms two right angled-triangles, and it is the

opposite side to both **angle B** and **angle C.**

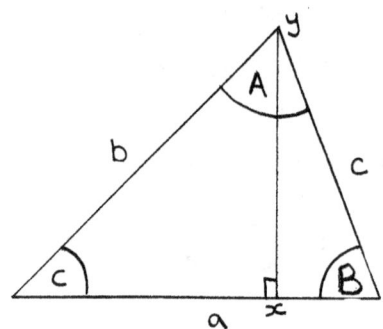

Fig 4-8b Height of triangle

The formula; $\sin = \dfrac{Opp}{hyp}$ can be used as; **opp = sin x hyp .**

opp = sinB x hyp .... opp = sin75 x 56.739.............. opp = 54.8

opp = sinC x hyp .... opp = sin40 x 85.264.............. opp = 54.8

The height of the triangle shown by **xy** is..................................... **54.8.**

The fact that the two equations involving the base angle from each triangle produces the same

answer is a good way to check that all calculations are correct, as the opposite side is shared by

both triangles.

4c Basic Geometry of the Circle.

The geometry of a circle is used extensively in the manufacture of all round constructions such as

pressure vessels, heat exchangers, etc.

A perpendicular line to a chord, if passing through the centre, will bisect that chord. **Fig 4-1c.**

AB is a chord.

CD is a perpendicular to that chord.

CD passes through centre O.

AC = CB.

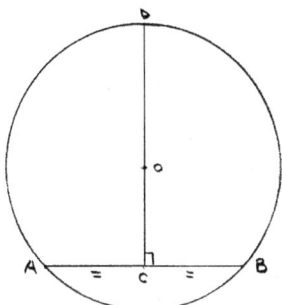

Fig 4-1c Perpendicular to chord

Example.

Fig 4-2c. A manhole on a tank is 800 diameter; this size will be the chord.

If the tank is 3000 diameter, it can be seen that the equation;

$\sin = \dfrac{opp}{hyp}$ will produce the angle needed to pinpoint the outside of the manhole on the tank.....

$\sin = \dfrac{400}{1500}$ .... sin = 0.266666666...

angle = 15.466°.

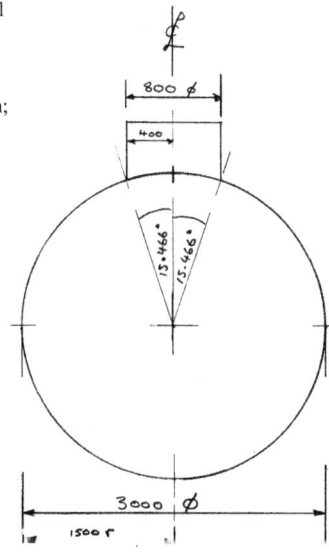

Fig 4-2c Example of bisected chord

If a triangle bounded by a circle uses the diameter as one of its sides, then the angle opposite the diameter will be 90°. **Fig 4-3c. AB is the diameter. C° = 90°.**

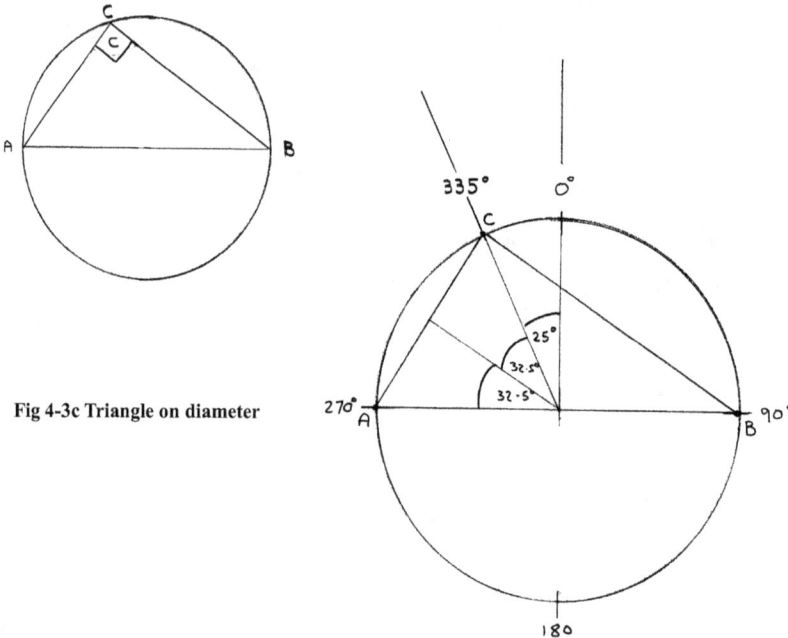

Fig 4-3c Triangle on diameter

Fig 4-4c Finding a point inside a vessel

This information can be used to locate a point on the inside of a vessel under construction.

Example, fig 4-4c. A point 25° away from the zero line of a vessel is to be marked on the inside.

It is good practice to always mark **0°, 90°, 180°** and **270°** on both the inside and outside of plates to be used for large vessels before rolling as seen in chapter 3.

The length of the chord **AC** can be determined by

90 minus 25 equals 65, which divided by 2 equals 32.5°

If the **inside diameter** of the vessel is **3000** then; **opp = sin x hyp......**

opp = sin 32.5 x 1500 = 805.9

which can be rounded up to 806, and when multiplied by 2 will give the length of the chord **AC** to be 1612.

The length of the chord **BC** can now be calculated to be $a^2 = c^2 - b^2$ .... $a^2 = 3000^2 - 1612^2$

$a^2 = 6401456$........... $\sqrt{6401456} = $ **2530.**

Point **C** can be marked from **A** and **B** either physically or be means of laser digital tape.

Triangles bounded by a circle sharing a chord as a base will have equal apex angles at the point where they meet the circumference. **Fig 4-5c.**

angle C = angle D = angle E.

angle O = angle C x 2.

The apex angle of the **same chord** when it is at the **centre** of the circle will be double the value of the circumference angle.

This information can be used when marking out the inside of large circular constructions.

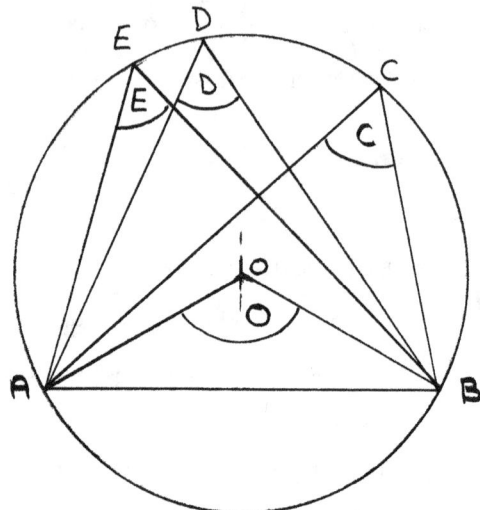

Fig 4-5c Triangles bounded by a circle

Example. The inside of a large circular construction requires that 5 centres be marked for pipes to be fitted at a later date. The only information given is to be found in the plan view **fig 4-6c**. **The first thing to do is convert the sizes given into degrees of the circle.**

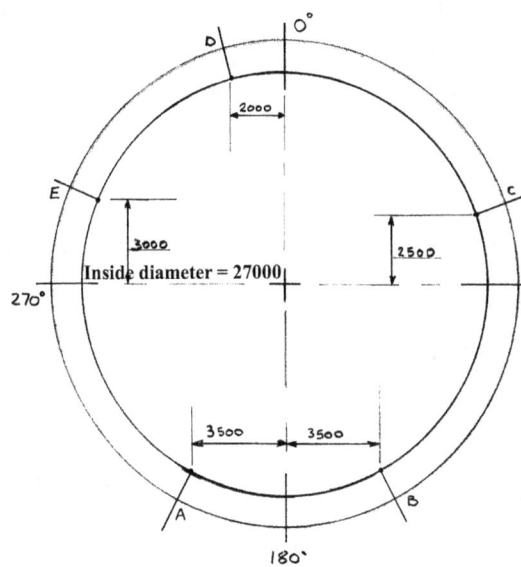

The diameter is 27000, therefore the radius is half of this value

.... **13500.**

$$\sin = \frac{opp}{hyp}$$ .... Angle at **A =**

$$\sin = \frac{3500}{13500}$$ ....**15.026°**

$180 + 15.026 = 195.026$

.... **A is at**

195.026°.

Therefore B will be

$180 - 15.026 = 164.974...$ **B is at**

164.974°.

Fig 4-6c Multiple points inside a circle

The angle at **C =** $\sin = \dfrac{2500}{13500}$ .... **10.67°.**

$90 - 10.67 = 79.33$...................................... **C is at 79.33°.**

The angle at **D =** $\sin = \dfrac{2000}{13500}$ .... **8.519°.**

$360 - 8.519 = 351.48$................................. **D is at 351.48°.**

The angle at **E** − $\sin = \dfrac{3000}{13500}$ .... **12.839°.**

$270 + 12.839 = 282.839$............................. **E is at 282.839°.**

The degrees of the circle are shown in Fig 4-7c.

The length of all the chords can now be

determined:

The chord **AB** can be seen to be **7000**.

This can be verified as follows;

195.026 - 164.974 = 30.052. The angle at the

centre using the chord **AB is 30.052°.**

It can be seen that this is an Isosceles triangle.

Therefore **30.052** is divided by **2** to give **15.026°**

which is the angle opposite half of the chord **AB**

in a right-angled triangle with the radius used as the hypotenuse.

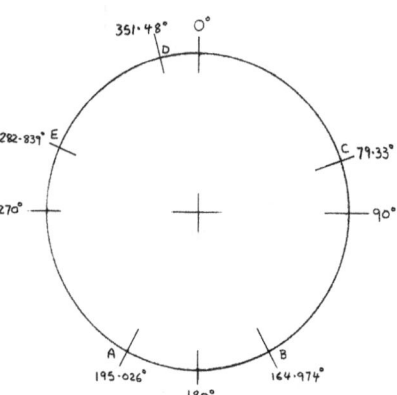

Fig 4-7c Conversion to degrees

opp = sin x hyp........... opp = sin 15.026 x 13500 .... = **3499.74** which is multiplied by 2 to

give the answer **6999.48** which is rounded up to **7000.**

The chord AB is 7000.

The angle at the centre of the chord **BC** is..... **164.974 - 79.33 = 85.644°.**

85.644 divided by **2 = 42.822°** .... opp = sin 42.822 x 13500..... = 9176.26..... x 2 = 18352.5

which is rounded up to **18353**

The chord BC is 18353.

The angle at the centre of chord **CD** is **(360 - 351.48) + 79.33** = ... **8.52 + 79.33 = 87.85°.**

87.85 divided by **2 = 43.925°**........ opp = sin 43.925 x 13500 ... = 9365.168.... x 2 = 18730.3

which is rounded down to **18730.**

The chord CD is 18730.

The angle at the centre of chord **DE** is........ **351.48 - 282.839 = 68.641°.**

68.641 divided by 2 = **34.3205°**..... opp = sin 34.3205 x 13500.... = 7611.59.... x 2 = **15223.18**
which is rounded down to **15223.**

The chord DE is 15223.

The angle at the centre of chord **EA** is .... **282.839 - 195.026 = 87.813°.**

87.813 divided by 2 = **43.906°** .... opp = sin43.906 x 13500.... = 9361.943.... x 2 = **18723.88**
which is rounded up to **18724.**

The chord EA is 18724.

It will be necessary to determine the chord **AD** also.

The angle at the centre of chord **AD** is .... **351.48 - 195.026 = 156.454°.**

156.454 divided by 2 = **78.227°**..... opp = sin 78.227 x 13500..... = 13216.0.... x 2 = **26432.**

The chord is 26432.

Every chord can be determined by this method, or, by dealing with three separate triangles to find the missing sides and angles using those chords now known.

All triangles using the chord AB will have the angle 15.026° opposite the chord.

Fig 4-8c shows the three triangles needed to mark **all points** from **A** or **B**.

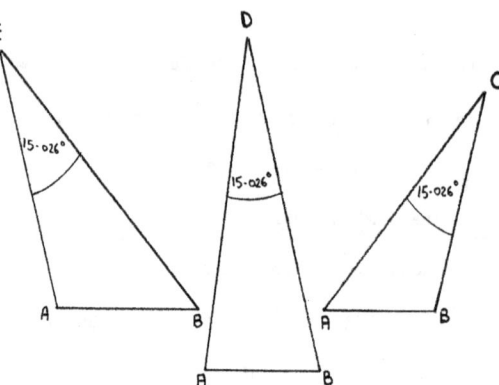

Fig 4-8c Three triangles needed to determine all values

Every angle and side can be determined using the information given in **section 4-b**.

Once the value of all necessary chords and angles are known they can be cross-referenced for accuracy when marking points **C, D and E.**

At this stage it is necessary to mark points **A** and **B, on** each side of **180°** as starting points.

Referring back to **Fig 4-7c.**

Half the chord between **A** and **B** has a centre angle of **15.026°.**

15.026 divided by 2 = **7.513°..... opp = sin 7.531 x 13500.... = 1769.34 .... x 2 = 3538.69** which is rounded up to **3539.**

This chord is marked directly from **180°** to give points **A** and **B.**

The values of the **radius, diameter and chords** can be

determined if two values are known:

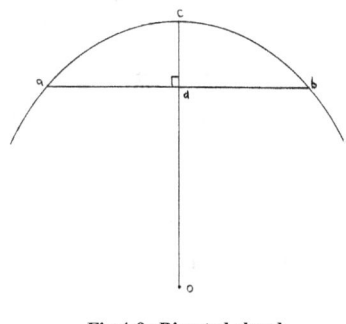

Fig 4-9c. The chord **ab** has been bisected

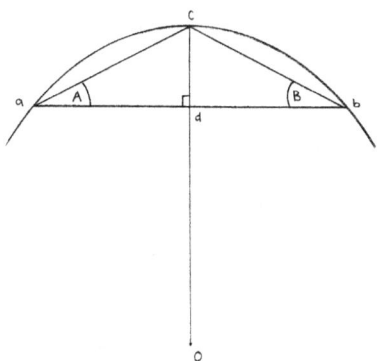

Fig 4-9c Bisected chord

Fig 4-10c Identical right-angle triangles

Thereby creating two identical right-angled triangles;

adc, and **bdc. angle A = angle B,** therefore the two

angles at point **c** are equal if the angles at **d** are all

90°. Fig 4-10c.

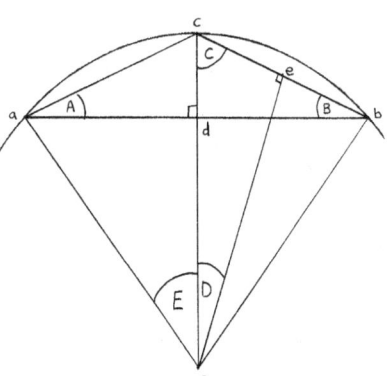

Fig 4-11c Triangles with common sides and angles

If the lines **oa** and **ob** are drawn two more identical right-angled triangles are created; **oda** and

odb. When the chord **cb** is bisected at point **e** with a perpendicular from centre point **o,** the right-

angled triangle **oec** is created.

Fig 4-11c. It can be seen that angle **B is equal to angle D** because **oec** and **bdc** are both right-angled triangles with a common angle **C**. It can also be seen that angle **E** will be equal to twice

the value of angle **D,** since the triangle **ocb** is actually two identical right-angled triangles sharing

the common side **oe.** It can further be seen that the value of the line **od** is **oa** minus **dc,** because

oa and **oc** are both radii and therefore have the same length.

Example. Fig 4-12c.

ab = 288.   dc = 32.

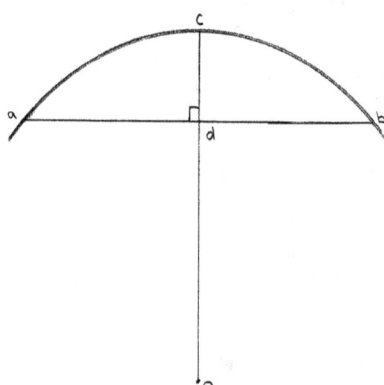

Fig 4-12c Two sides known

solution.......

$$ad = \frac{288}{2} = 144$$

This example should be dealt with using **fig 4-11c** as a reference.

ac =.............ac²= ad²+dc².......ac²= [144²+32²]...........ac = √21760 ....ac = 147.512.

ac and bc are equal. The mid-point of **bc** is at point **e.**

$$ce = \frac{147.512}{2} \text{ ....ce = 73.756}$$

angle B =........ $\sin = \dfrac{opp}{hyp}$ **.......** $\sin B= \dfrac{32}{147.512}$ **.......sin B = 0.216931503.......angle B = 12.5288°**

angle C = 90° _minus_ 12.5288°........... angle C = 77.4712°

oe = tan C x adj............... oe = tan C x 73.756..........oe = 331.9

$oc^2 = oe^2 + ce^2$.............$oc^2 = [\ 331.9^2 + 73.756^2\ ]$...............$oc = \sqrt{115597.557}$ ....oc = 339.996

od = oc-dc............od = [339.996 - 32]...............od = 307.996

angle D = $\sin = \dfrac{opp}{hyp}$ .... $\sin D = \dfrac{73.756}{339.996}$ ....sin D = 0.216931963......angle D = 12.5288°

which is the same as **angle B**

angle E = D x 2.................angle E = 12.588 x2...............angle E = 25.0576°

The diameter of a circle is the radius multiplied by 2. Since we know the radius of this circle to

be ...... oe = 339.996, then the diameter is 339.996 x 2 = ... 679.992

The angle of the arc between **a** and **b** is E x 2........ **angle of arc = 25.0576° x 2 = ...angle of**

arc = 50.1152°.

Length of arc = [π x diameter] divided by 360 for the number of degrees in a full circle

multiplied by the angle of arc .... π x 679.992 – 2136.2578..... $\dfrac{2136.2578}{360}$ = 5.934... this is the length of arc for one degree. **length of arc ab = 5.934 x 50.1152 .... = 297.38359°.**

4d Geometry of regular Polygons.

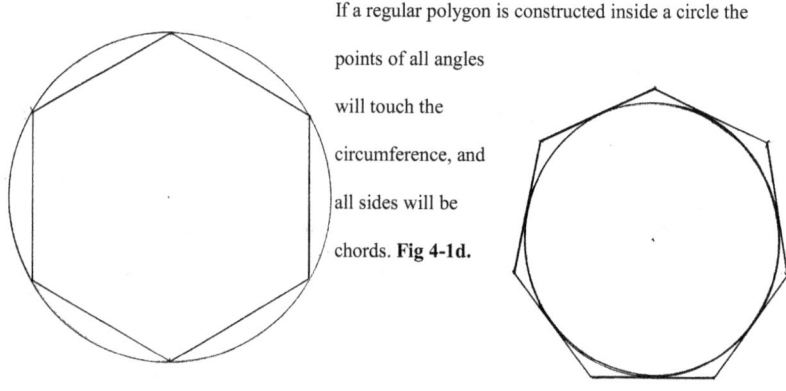

If a regular polygon is constructed inside a circle the points of all angles will touch the circumference, and all sides will be chords. **Fig 4-1d.**

Fig 4-1d Regular polygon inside a circle

Fig 4-2d Regular polygon outside a circle

If a regular polygon is constructed outside a circle all sides will touch the circumference at their mid points, and all sides will be tangents. **Fig 4-2d.**

Regular polygons can be constructed by dividing the number of sides into 360, which is the number of degrees in a full circle, then all relevant information can be obtained if the length of polygon side is known.

Example... **Fig 4-3d.** A regular **5** sided polygon or pentagon has a side length of **42.**

$\frac{360}{5} = 72$ The angle at centre **O** of each of the 5 triangles is **72°.**

The remaining two angles are equal and are found by subtracting 72 from 180 and dividing the

answer by 2 to produce **54°.** A right angled triangle is obtained by bisecting the side **ab** with a

perpendicular from the centre at **O** producing

point **c.**

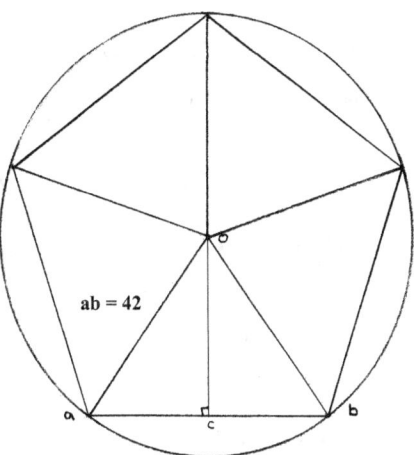

The length of side **ac** in the triangle **aoc** is

$\frac{42}{2} = 21$ the angle opposite **ac** is $\frac{72}{2} = 36$ the

third angle is **90 - 36 = 54** as we have seen.

Since all angles are known and the length of

the shortest side is known to be 21......

$hyp = \frac{21}{sin36}$ .... **hyp = 35.72.** This will be

the radius of the circle bounding the polygon.

ab = 42

**Fig 4-3d Regular polygon with length of side
given**

CHAPTER 5

Pyramids and cones

5 a True length.

All pyramids and cones are easier to mark out if its **apex** can be used. The examples in this chapter will all be marked out onto the template paper or directly onto the material using the apex.

When working directly from drawings, although all dimensions given will be correct, **true length** will often need to be determined because working drawings mostly show **apparent length**.

The difference can be explained with a simple example:

A person standing directly in front of a ladder leaning against a wall will see its **apparent length.**

If the person moves to a view point at the side where the ladder can be seen leaning at an angle, its **true length** will now be seen as the hypotenuse of a right-angled triangle, **fig 5-1a.**

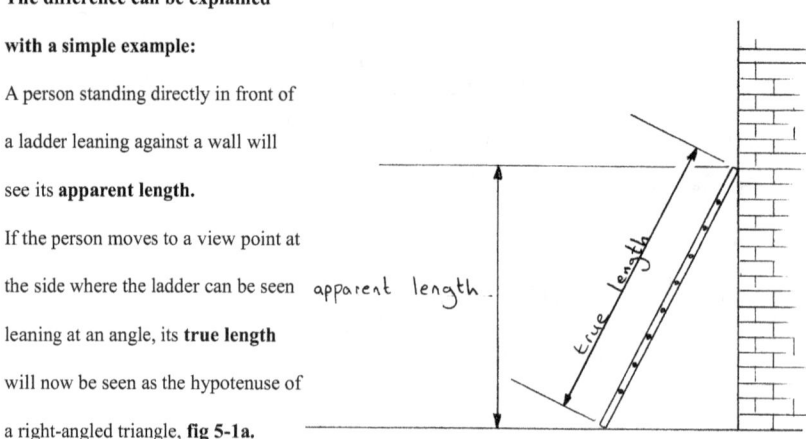

Fig 5-1a True length

Example.

To achieve the correct view point in the workshop the drawing is rotated by using a pair of dividers as in **fig 5-2a**, where the line **AB** is shown in two views; the side elevation and the plan, both of which show its **apparent length.**

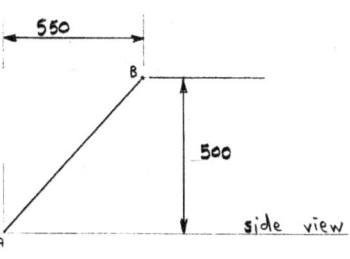

In this example point A is used as the pivot.

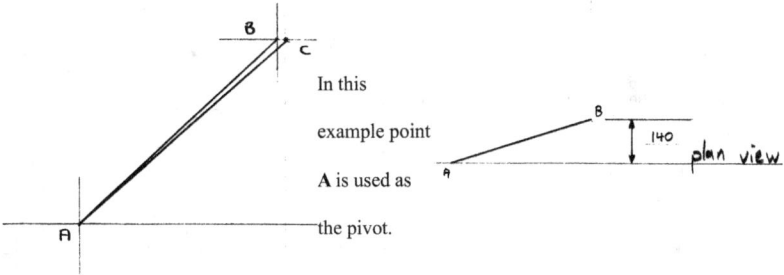

Fig 5-2a Apparent length

Fig 5-3a. Set the dividers to the length **AB** on the plan view and scribe the arc down to the base line. At the intersection with the base line draw a perpendicular up to the side view to intersect with the height line to produce point **C.**

The line **AC** is the true length of the line **AB.**

Fig 5-3a True length

The true length can also be found using trigonometry as follows:

Firstly determine the length of **AB** in the plan view.

$a^2 + b^2 = c^2$ .... $550^2 + 140^2 = 322100$ .... $\sqrt{322100} = 567.538$

Using this value as the new horizontal size in the side view, the true length can now be found.

$a^2 + b^2 = c^2$ .... $567.538^2 + 500^2 = 572099$ .... $=\sqrt{572099} = 756.372$.

The true length of **AB** is **756.372.**

5b Regular square pyramid.

A pyramid has a base with four sides which taper vertically to an apex point. It can be cut off at any height to produce an opening. If the apex of the pyramid is directly above its base, it is said to be a regular pyramid.

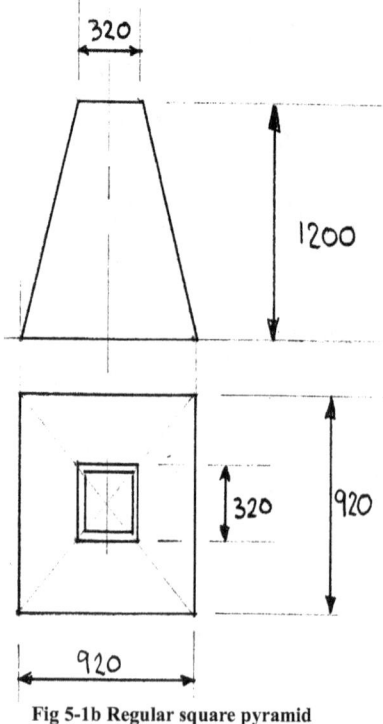

Example, fig 5-1b, a square pyramid has a base of 920 x 920 od tapering to 320 x 320 od at a height of 1200, with the thickness of the plate being 10.

Fig 5-1b Regular square pyramid

When bending plate with square corners, only inside sizes are used, therefore the base can be considered to be 900 x 900 and the small end to be 300 x 300 when marking the layout.

Fig 5-2b. The apex angle of this pyramid can be located by drawing a base line 900 long, then bisect it to find the centre and draw a perpendicular.

At a height of 1200 mark a distance of 150 on either side of a line parallel to the base line.

The apex can then be located by continuing the sides to the intersection point on the centre line.

On the **plan view** the centre point **O** is used as the **pivot point** to project the points **B** and **1** up to the centre line, where they can be projected vertically to give points **B** and **1** in the **side view**, giving the **true length; B1** needed for the development of the template, which works well on a small scale, but accuracy on larger work dictates that that the apex be determined mathematically:

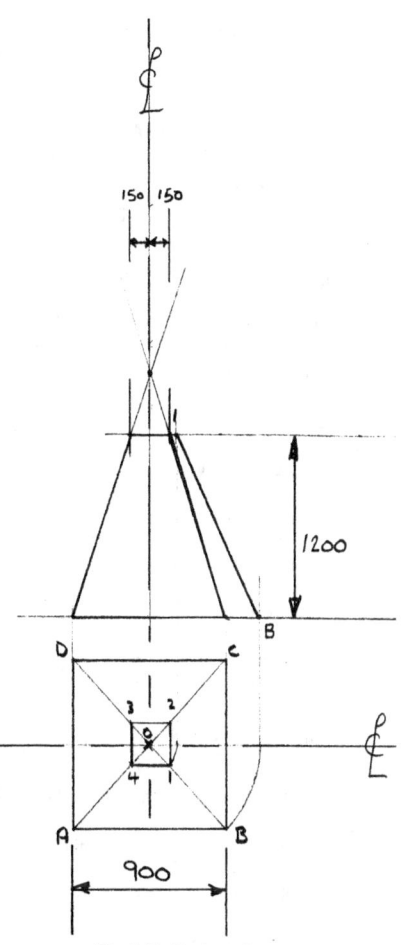

Fig 5-2b Projected apex

Example.

This is the same pyramid as in fig 5-1b and fig 5-2b.

To locate the apex, the pyramid angle must be determined as follows, **fig 5-3b.**

450 - 150 = 300 .... which, is used as the base of a right-angled triangle, the height of which is 1200.

With these two pieces of information we can determine the pyramid angle to be...........

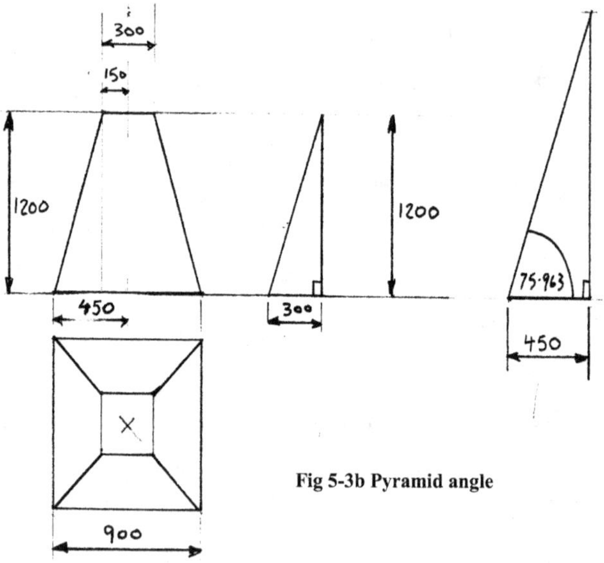

Fig 5-3b Pyramid angle

$\tan = \frac{opp}{adj}$ .... $\tan = \frac{1200}{300}$ ...... tan = 4........ tan 4 = 75.963°

Knowing the angle of the pyramid to be 75.963° the apex height can be determined as;

opp = tan x adj .... opp = tan 75.963° x 450 .... opp = 4 x 450 = 1800. With the knowledge that the apex height of this pyramid is 1800 the layout can be completed:

In **fig 5-4b** the apex is marked correctly and points **B** and **1** are projected vertically up to the side view as in **fig 5-2b.**

In the side view use the apex centre **O** to describe a greater arc from point **B** and a lesser arc from point **1**. Draw a line from any point on greater arc **B** back to the apex centre **O**. This new point will be **x.**

With the dividers set to the distance **AB** on the plan view, use point **x** on arc **B** to step off 4 equally spaced points on the arc.

A line drawn from each of these points back to the apex centre **O** is a bending line.

A line drawn to connect each point to its neighbour as a chord is a cutting line.

The process can be repeated on the lesser arc **1** with the dividers set to the distance **1-2** on the plan view, where it will be seen that the four points meet the four radial lines.

Draw in the four chords on the lesser arc, which are also cutting lines, to complete the template.

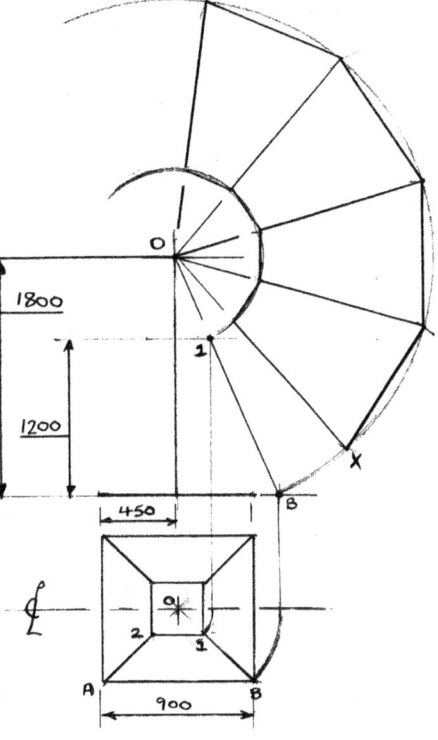

Fig 5-4b Square pyramid template

The template could not have been drawn without using the **true length** of the diagonal line between the apex centre and one of the bottom corners, to produce the greater arc.

This length can be found mathematically allowing for the development to be drawn directly onto either template paper or the material to be used for the job:

In **fig 5-4b** on the plan view the distances **OB** and **O1** were transferred to the plan view to give the greater and lesser arc values.

It can be seen on the plan view that **OB** is the hypotenuse of a right-

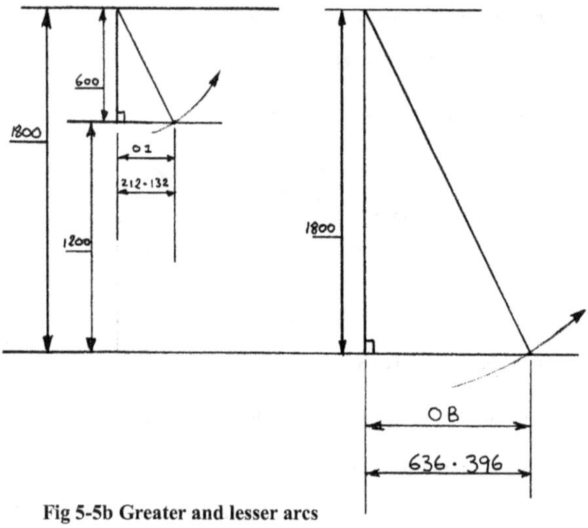

Fig 5-5b Greater and lesser arcs

angled triangle with the other two sides both being 450 long. Therefore $a^2 + b^2 = c^2$....

$450^2 + 450^2 = 405000$.... $\sqrt{405000} = 636.396$.

OB = 636.396.

It can also be seen that **O1** is the hypotenuse of a right-angled triangle with the other two sides both being 150 long. Therefore $150^2 + 150^2 = 45000$ .... $\sqrt{45000} = 212.132$.

O1 = 212.132.

In **fig 5-5b**, These two values are used to determine the radius sizes of the greater and lesser arcs :The apex height is known to be 1800 which is used as one side of a right-angled triangle and **OB** is known to be 636.396 which is used as the second side of the right-angled triangle ....

$a^2 + b^2 = c^2$ .... $1800^2 + 636.396^2 = 3644999$ .... $\sqrt{3644999} = 1909.188$.

The radius size for the greater arc is 1909.

For the lesser arc the apex height is .... 1800 - 1200 = 600 which again is one side of a right-

angled triangle with the value **O1** (212.132) being the

second side...

$600^2 + 212.132^2 = 404999$ .... $\sqrt{404999} = 636.39$.

The radius size for the lesser arc is 636.39.

Once these two arcs have been drawn four chords 900 long

are stepped off on the greater arc and the points are all

drawn back to the radius centre as radial lines. The four

chords 300 long marked on the lesser arc will intersect with

the radial lines, and with all chord lines drawn in the

development is complete.

5c Regular square pyramid with sectional cut.

Example.

In **fig 5-1c**, the pyramid shown is much the same as the

previous example with the exception of the inclined cut, but

it is marked out in exactly the same way ....

Since only inside sizes are used it will be assumed that

those shown are inside sizes.

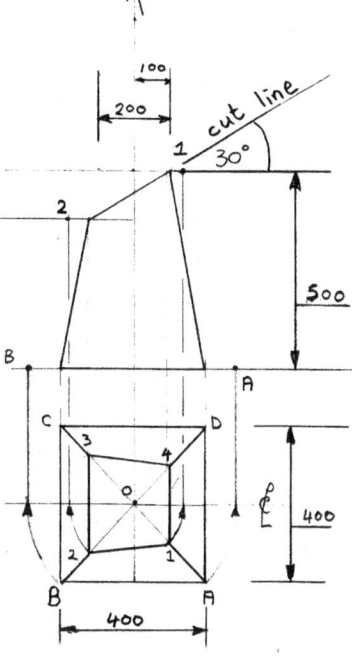

Fig 5-1c Pyramid with sectional cut

The pyramid has a 400 x 400 base, is 200 x 200 at the 500 high mark with a 30° cut at that level.

On the plan view with **O** as the pivot, **OA** and **OB** have been projected to the side view base line.

O1 has been projected to the top of the cut line on the same side as **A** to give the true length of the line **A1.**

O2 has been projected to the bottom of the cut line on the same side as **B** to give the true length of the line **B2.**

Calculate the apex height as in the previous exercise by first determining the pyramid angle

....

$200 - 100 = 100$ .... $\tan = \dfrac{opp}{adj}$ .... $\tan = \dfrac{500}{100}$ .... $\tan = 5$ ... = angle 78.6 °

True length of OA and OB =... $a^2 + b^2 = c^2$.... $200^2 + 200^2 = 80000$ ... $\sqrt{80000} = 282,84$

Knowing the pyramid angle to be 78.6°, this is used to find the apex height by using it in the formula:

$opp = \tan \times adj$ .... $opp = \tan 78.6° \times 200$........ $opp = 4.9594 \times 200 = 991,889.$

Apex height is 991.889........... this is rounded up to **992.**

In **fig 5-2c** the apex height is marked as the centre point **O** for the development.

Point **A** is marked on the side view base line at 282.84 from the centre line, it can be seen that a distance of 282.84 from the centre line will also produce point **B.**

Points **1** and **2** are marked on the side view by projection as in the last exercise.

Points **A, 1** and **2** are used to draw arcs with point **O** at the centre.

With the dividers set at 400 which is the length of **AB,** from point **A** on the greatest arc step off four chords, labelled **A -B, B-C, C-D , D- A.** The reason **A** is used twice is because it is both the beginning and the end of the development.

Draw in the chords and also the radial lines back to **O.**

It can be seen on the plan view that points **1** and **4** are at the highest point on the pyramid and a line drawn from the centre passes through point **1** ending at point **A**, likewise a line through point **4** ends at point **D.**

This means that on the development radial lines from **A** and **D** will intersect with the arc drawn from **1.**

In the same manner it can be seen that radial lines from **B** and **C** will intersect with the arc drawn from **2.**

Draw in the lines from each intersection point to its

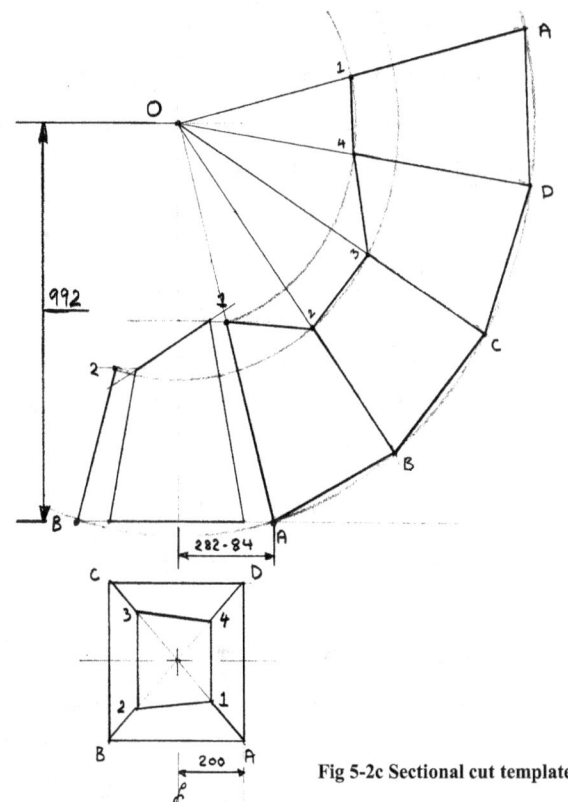

Fig 5-2c **Sectional cut template**

neighbour, which are in fact now cut lines as are the chord lines on the greater arc, with the radial lines being bend lines.

5d Oblique pyramid.

The word oblique means slanting, so an oblique pyramid will look off-centre, but, as will be seen a centre is still needed to make a template. As before, it will be assumed that sizes given are inside sizes.

In practice, in the workshop it is usually necessary to make one's own notes or working drawings from those supplied. As in this case only inside sizes are needed, they may have to be determined from the drawings supplied and transferred to the boilermakers own, more specific drawing.

Example.

Fig 5-1d shows an oblique pyramid with a base 400 x 400 and an opening of 200 x 200 at a height of 450. The opening at the top is 200 off-set to the opening at the bottom. The angle, of the off-set centres of the pyramid, together with the inner and outer angles, are used to find the apex height, by means of converging lines.

Calculating the angles is far more accurate than simply continuing the lines of the drawing.

The off-set is 200; this is the base of a right-angled triangle.

The height is 450; this is the horizontal side of the triangle.

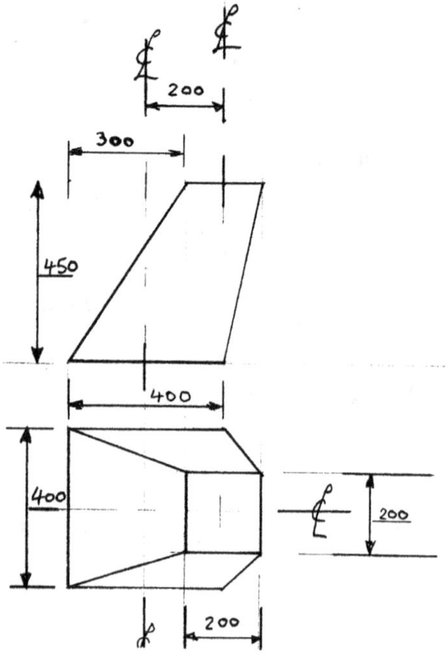

Fig 5-1d Oblique pyramid

$$\tan = \frac{opp}{adj} \;\ldots\ldots\; \tan = \frac{450}{200} \;\ldots\ldots\ldots\; \tan = 2.25 \;\ldots\ldots\ldots \text{centre angle} = 66°.$$

By the same method the inner angle is found to be ... $\frac{450}{100}$ .... tan = 4.5 .... angle =77.47°

And the outer angle is $\frac{450}{300}$ .... tan 1.5........... angle 56.3°

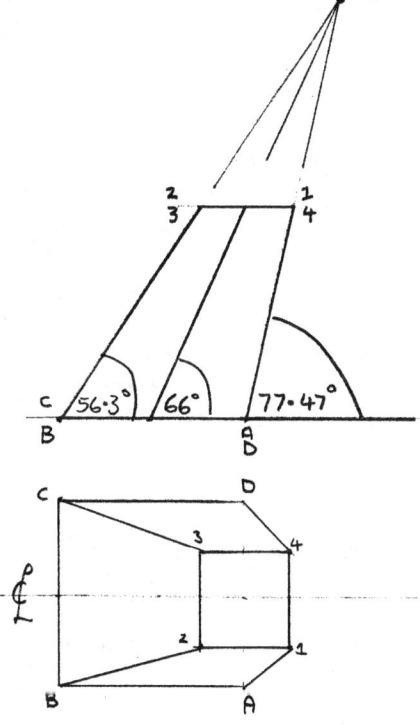

In **fig 5-2d,** the three angles are drawn to a convergence point at **O**, which, as in the previous examples is used as the centre point for all arcs in the template.

Fig 5-2d Converging angles

This centre is now projected down to the plan view, Fig 5-3d.

With the dividers set at **O**, bring points **A, B, 1** and **2** up to the centre line and project them vertically. These will be used as arc points from **O** for the template.

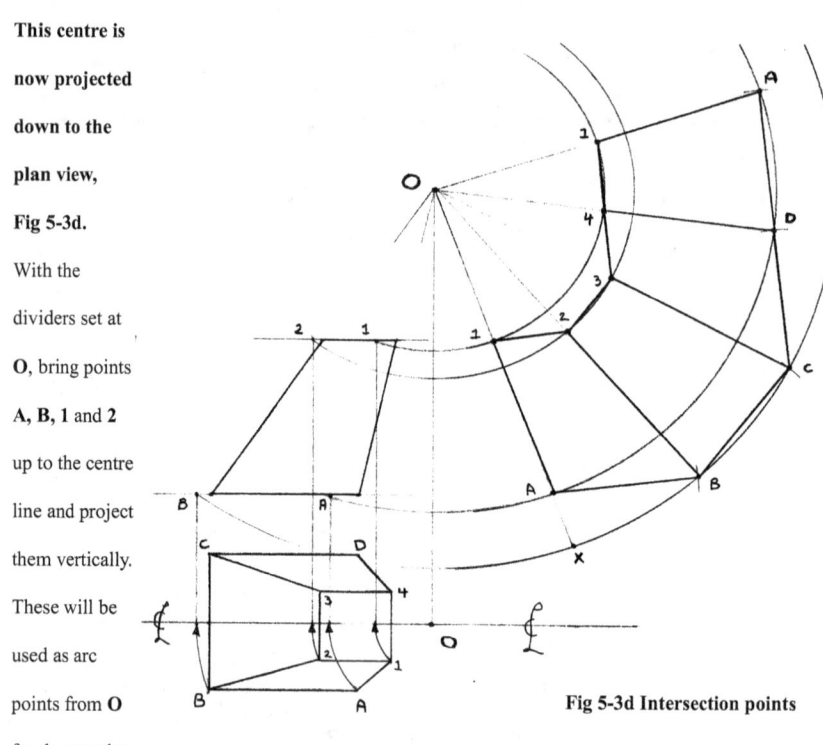

Fig 5-3d Intersection points

In **fig 5-3d,** the four arcs are drawn from **A, B, 1** and **2.**

From a point on the greatest arc, labelled point **X,** draw a radial line back to point **O.**

The intersection of this radial line OX with the arc drawn from A will give the starting point A.

With reference to **the plan view,**

set the dividers at **AB** on the plan view and draw an arc from **A,** to intersect arc **B** at a point

labelled **B**, and draw a radial line back to **O**.

With the dividers set at **BC** mark point **C** on arc **B,** from **B.**

With the dividers set at **CD** mark point **D** from **C** on arc **A,** and draw a radial line back to **O.**

With the dividers set at **DA** mark point **A** from **D** on arc **A,** and draw a radial line back to **O.**

Join all bottom points for the cutting lines.

Radial line **AO** intersects with arc **1** to give point **1.**

Radial line **BO** intersects with arc **2** to give point **2.**

Radial line **CO** intersects with arc **2** to give point **3.**

Radial line **DO** intersects with arc **1** to give point **4.**

Radial line **AO** intersects with line **1** to give point **1.**

Join all top points for the cutting lines.

The apex of this pyramid can also be found by using

the law of sine.

Example.

Fig 5-4d. Both the inner and outer pyramid angles are

known and the size of the base of the pyramid is also

known.

$$\frac{\sin A}{a} = \frac{\sin B}{b} = \frac{\sin C}{c}$$

The side opposite angle B is side b.

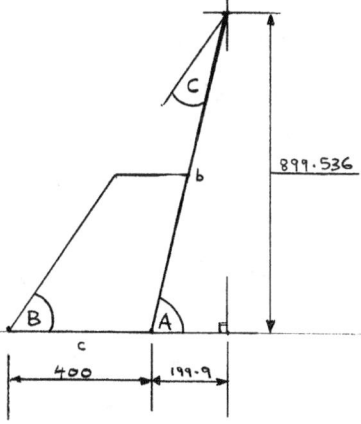

Fig 5-4d Apex found with sine law

The base will be side c, and its opposite angle will be C.

Angle C = angle A minus angle B......... 77.47 - 56.3 = 21.17. .... **Angle C is 21.17°.**

The formula will be converted to read ... $\dfrac{b}{\sin B} = \dfrac{c}{\sin C}$

$$b = c\left(\frac{\sin B}{\sin C}\right) \qquad .... \qquad b = 400\left(\frac{\sin 56.3}{\sin 21.17}\right) \qquad .... \; b = 921.484.$$

Knowing the bottom angle at **A** to be **77.47°** and the length of the hypotenuse, which in this case is side **b**, to be **921.484**, the opposite side is the apex height.

opp = sin x hyp.   opp = sin 77.47 x 921.484 = **899.536.**

The third side of the triangle is ... **c² = a² - b²** ...921.484² - 899.536² = 39967... √39967 = **199.9.**

400 + 199.9 = 599.9.

From point **B** at a distance of **599.9** a perpendicular is drawn, and at a height of **899.536** the **apex** for the pyramid can be marked.

It will be seen that the greater radius will be √599.9² + 899.536² .... = **1081.22.**

5e Regular cone.

A regular cone has a round base with its apex directly above the centre of its base.

It is one of the easiest templates a boilermaker will have to mark out, but the knowledge of its

construction is vital for the more complicated layouts that will be encountered.

The fact that the cone is round dictates that the **mean** size

must be used.

Example.

Fig 5-1e. A regular cone has a base diameter of 416 outside

diameter, with an upper size of 200 outside diameter, at a

height of 600, with the material thickness of 10.

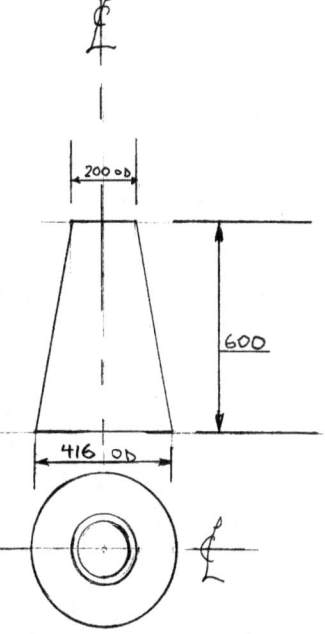

To determine the **mean** sizes needed the material thickness

must be **subtracted** from the outside sizes given:

base diameter is 416 - 10 = 406.

upper diameter is 200 - 10 = 190.

The cone angle is found by subtracting the upper radius

from the base radius the applying the tangent formula.

cone angle is 203 - 95 = 108...............

$\tan=\dfrac{opp}{adj}$ .... $\tan=\dfrac{600}{108}$ .... **angle = 79.796°.**

Fig 5-1e Regular cone

The apex height can now be determined using the base radius and the cone angle:

opp = tan x adj........ opp = tan79.796 x 203 .... opp = 1127.77.

The apex height is **1127.77** and the base radius is **203**. Therefore, the greater **arc radius** needed

is: $a^2 + b^2 = c^2$ .... $1127.7^2 + 203^2 = 1312.916$........ $\sqrt{1312.916} = $ **1145.82.**

The lesser **arc radius** is:

Apex minus 600 equals 527.77, and

the upper radius is 95.

$527.77^2 + 95^2 = 287566$

.......$\sqrt{287566}$........ = **536.25.**

This information is used to mark fig

5-2e with the base radius size of **203**

used as a chord length stepped off **6**

times along the greater arc, and the

upper radius size of **95** used as a

chord length stepped off **6** times

along the lesser arc. The template is

now complete because:

....

The radius of a circle will produce

6 equal chords on its

circumference.

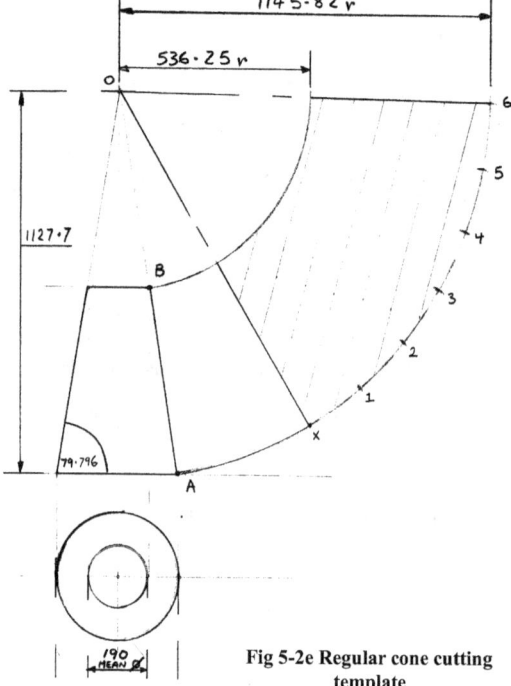

Fig 5-2e Regular cone cutting
template

5f. Regular cone with sectional cut.

Example.

Fig 5-1f. A regular cone has a **mean** base diameter of **500** with a **40°** cut at a height of **700** and a cone angle of **76°**.

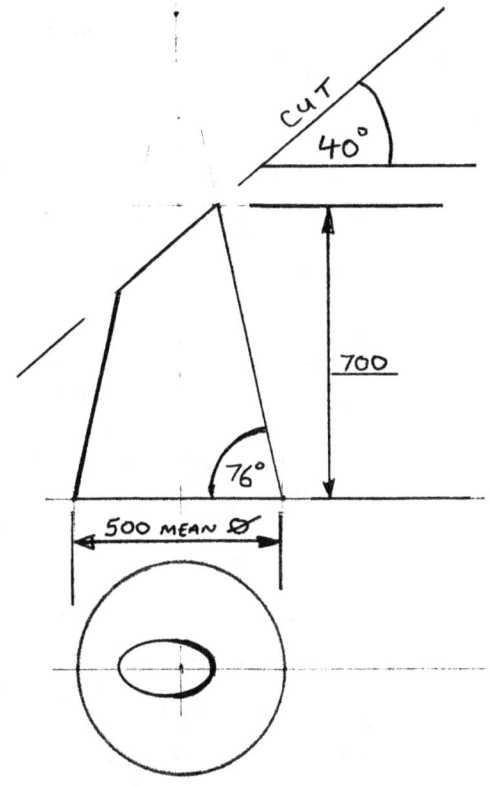

Fig 5-1f Regular cone with sectional cut

The **apex height** is; opp = tan x adj........ opp = tan76 x 250 .... **1002.69.**

The **greater arc** is; $1002.69^2 + 250^2 = 1067887.236$........ $\sqrt{106788.236} = $ **1033.38.**

This information is used to mark the **greater arc** in **Fig 5-2f.** All of the lesser arcs will need to be projected thus:

The circle of the base in the plan view is divided into equal parts. Large jobs will need many equal parts. **For this exercise only 6 will be used.**

These points on the plan view circumference are projected vertically to the base of the side view. Lines are then drawn in to converge at the **apex centre** producing intersection points with the **40° cut.** **These intersection points are used to draw in all of the lesser arcs.** The base **radius** from the plan view is used to mark **6 equal chords** on the greater arc and radial lines are then drawn back to the **apex centre** from these points.

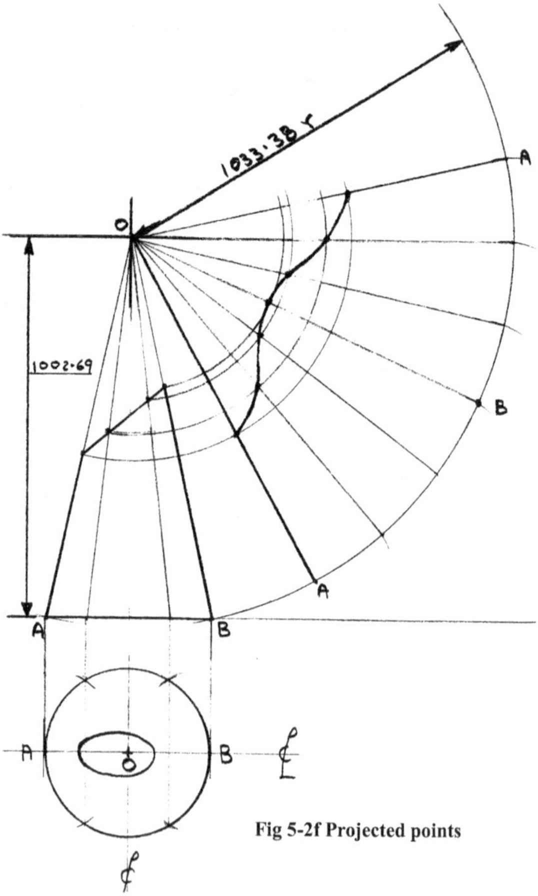

Fig 5-2f Projected points

As before point **A** will be both the starting and finishing point

for the template.

Point **B** will be the middle point of all the chords.

The diameter of the plan view is marked **AOB.**

It can be seen that lowest point of the cut intersects **AO**

and the highest point of the cut intersects **OB.**

Therefore.... On the template the **outer radial lines** marked

A will intersect with the **lowest arc**

and the **central radial line** marked **B** will intersect with the

highest arc.

The remaining intersection points follow in progression to

complete the template.

These points are now joined together with a fluid curved line.

This will be achieved with practice.

The template is now complete.

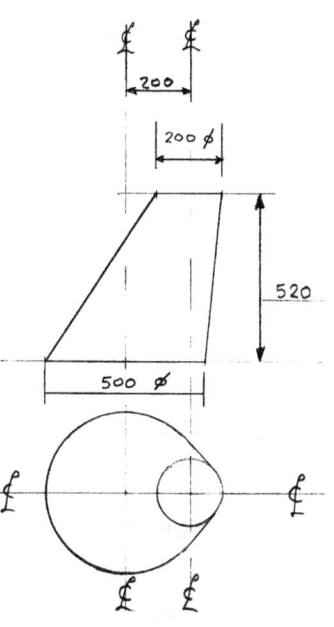

Fig 5-1g Oblique cone

5g. Oblique cone.

An oblique cone is one with its apex centre not directly above its base centre.

The apex centre however, will still be used to mark out the template.

Example. Fig 5-1g.

An oblique cone has a base **mean diameter** of **500** with an opening of **200 mean diameter** at a height of **520.**

As before the first thing to do is to locate the **apex centre.** The inner, outer and centre cone angles can be determined using the **tangent equation** as shown in **Fig 5-2g** where, all sizes have been shown for clarity.

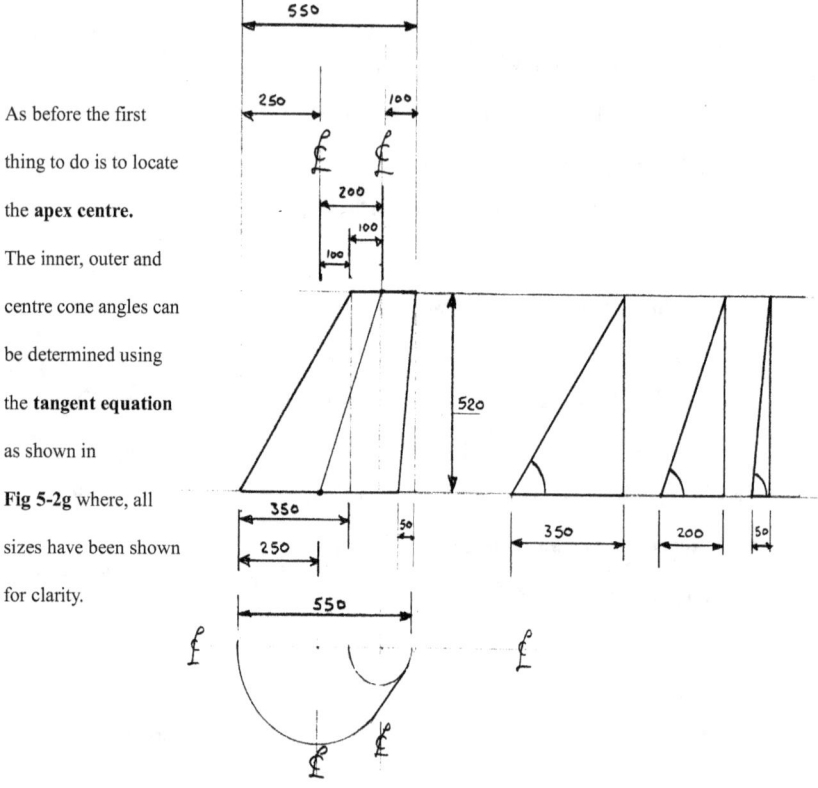

Fig 5-2g Cone angles

$\tan = \dfrac{opp}{adj}$ .... in all three cases the opposite side is **520**.

Inner angle.... adj is **50**........... angle is **84.5°**.

Centre angle... adj is **200** .... angle is **68.96°**.

Outer angle.... adj is **350** .... angle is **56.06°**.

Knowing these three angles and the base

diameter of **500** the apex position can be marked

as in the previous exercise, by either drawing in

the angles or mathematically determining its

position.

In **Fig 5-3g**, with the apex position marked **O**,

this is transferred down to the plan view centre

line.

The two circles in the plan view are divided into

6 equal parts.

Dividing a semi-circle into 3 equal parts will

produce the same result.

The points on the large base circle will be

labelled **A, B, C** and **D**.

The points on the smaller upper circle will be

labelled **1, 2, 3** and **4,**

Using point **O** as a centre all points on both

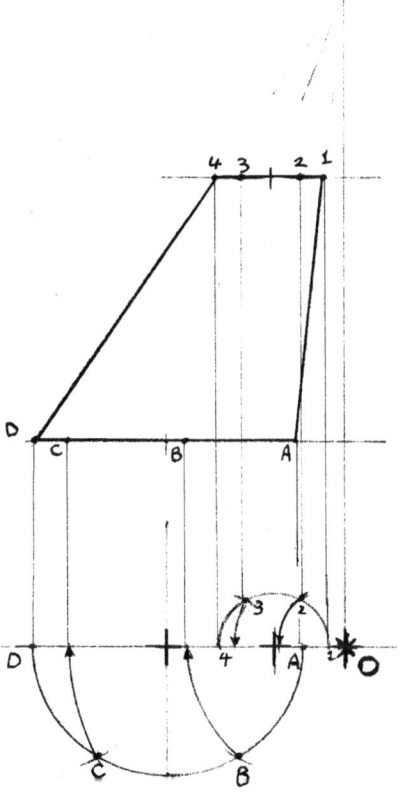

Fig 5-3g Apex projected to plan view

circles are brought up to the plan view centre line then projected vertically to the side view.

These will be used as arc points from **O** for the template.

In **Fig 5-4g** all arc lines are drawn and a radial line is drawn back to **O**.

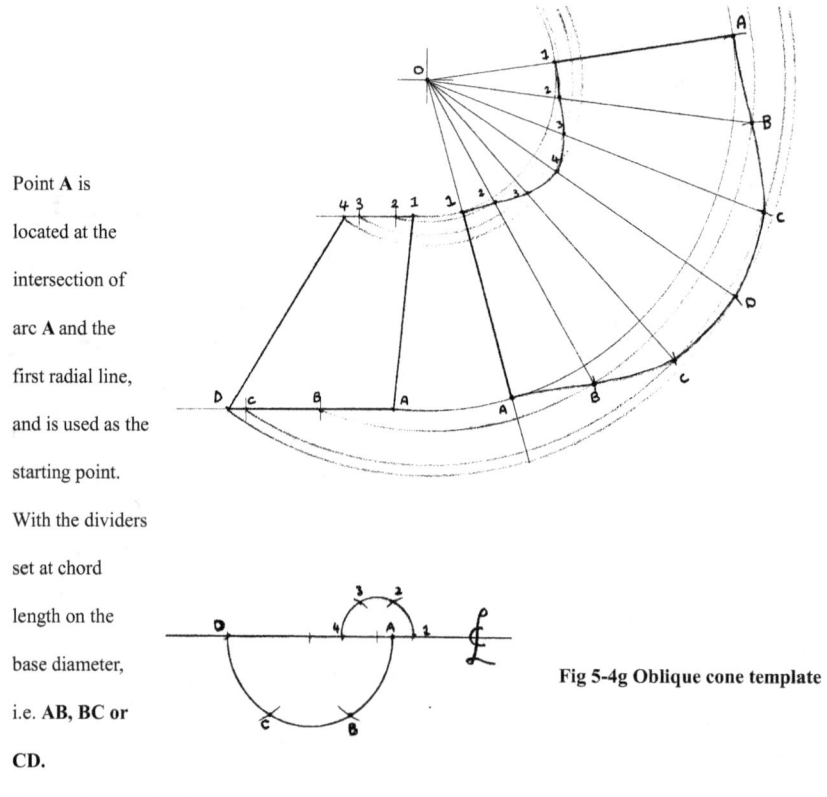

Point **A** is located at the intersection of arc **A** and the first radial line, and is used as the starting point. With the dividers set at chord length on the base diameter, i.e. **AB, BC or CD.**

Fig 5-4g Oblique cone template

From point **A** mark point **B** on arc **B.**

From point **B** mark point **C** on arc **C.**

From point **C** mark point **D** on arc **D.**

At this point the template goes into reverse.

From point **D** mark point **C** on arc **C.**

From point **C** mark point **B** on arc **B.**

From point **B** mark point **A** on arc **A.**

Radial lines drawn back to **O** from these points will produce intersection points on arcs **1 - 4.**

These intersections, line up with the chord lengths on the small diameter; 1-**2, 2-3 and 3-4.**

A fluid line is then drawn connecting all lower points.

Another fluid line is drawn connecting all upper points.

Both of these lines are cutting lines.......... The template is now complete...

CHAPTER 6

Transitions

A transition is an intermediate section between one shape and another assisting with the smooth flow of whatever product is being transferred. An example would be a square to round at the bottom of a square hopper guiding material into a round pipe.

6a Rectangle to square.

Example.

Fig 6-1a shows a rectangle of **500 x 400** which becomes **200 x 200** square at a height of **300,** which is made in two halves.

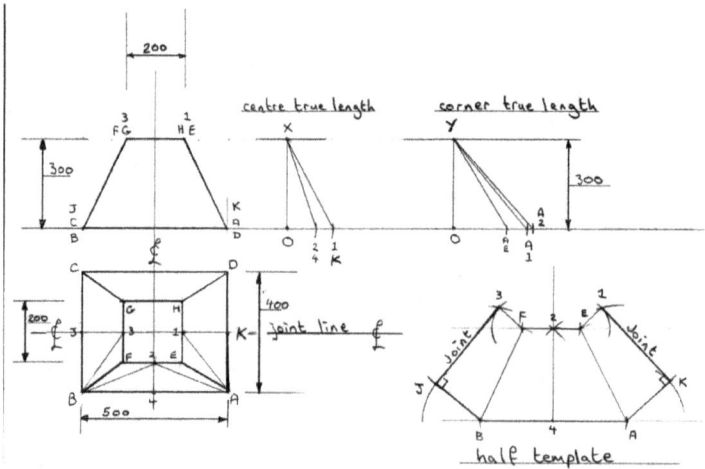

Fig 6-1a Rectangle to square

© The Mathematics of Boilermaking by Jim Draper 2017
76

Step 1 is to determine **all true lengths** needed and since the figure is symmetrical the sizes can **all** be found using only one corner:

A perpendicular line **OX** is drawn to the **same height** as the transition which is **300.**

With the dividers or trammel points set at the distance **2-4** on the plan view, mark a point from **O** and draw a line back to point **X** which will be the **true length** of the line **2-4.**

Using the distance **1K** on the plan view, mark a point from **O** and draw a line back to point **X** which will be the true length of the line **1K.**

A perpendicular line **OY** is drawn to the **same height** as the transition which is **300.**

With the dividers or trammel points set at the distance **A1** on the plan view, mark a point from **O** and draw a line back to point **Y** which will be the **true length** of the line **A1.**

Using the distance **A2** on the plan view, mark a point from **O** and draw a line back to point **Y** which will be the **true length** of the line **A2.**

Using the distance **AE** on the plan view, mark a point from **O** and draw a line back to point **Y** which will be the **true length** of the line **AE.**

The half template can now be marked:

A line **AB** is drawn **500** long.

With the dividers or trammel points set at the **true length A2**, an intersecting arc is drawn from points **A and B** to produce point **2.**

With the dividers or trammel points set at the distance **E2** on the plan view (which is the same as **F2**) an arc is drawn on either side from point **2.**

With the dividers or trammel points set at the **true length AE** two arcs are drawn from points **A and B** to intersect at points **E and F.**

With the dividers or trammel points set at the distance **E1** on the plan view (which is the same as **F3**) an arc is drawn from points **E and F.**

With the dividers or trammel points set at the **true length A1** an arc is drawn from points **A and B** to intersect at points **1 and 3.**

With the dividers or trammel points set at the distance **AK** on the plan view (which is the same as **BJ**) an arc is drawn from points **A and B.**

With the dividers set at the **true length A1** an arc is drawn from points **1 and 3** to intersect at points **K and J. Since the angles at J and K are right angles in the plan view, they must be right angles in the finished template.**

This template will produce half of the transition piece a mirror image is required to produce the other half.

6b. Regular square to round.

Example.

Fig 6-1b shows a **400 x 400 square** end which becomes a **300 diameter** at a height of **400,** which is made in two halves. **Remember for round sections the mean must be used.**

As before **true lengths** will need to be determined and since the transition is symmetrical the sizes can be found using only one corner.

On the plan view the **diameter** is divided into **12** equal parts.

Since one corner only is required, a quarter of the circle between **E, A and F** is divided into **3** equal parts.

A perpendicular line **OX** is drawn to the **same height** as the transition which is **400.**

With the dividers or trammel points set at the distance **A1** (which is the same as **A4**) mark a point from **O** and draw a line back to **X** which will be the **true length** of the lines **A1 and A4.**

With the dividers or trammel points set at the distance **A2** (which is the same as **A3**) mark a point from **O** and draw a line back to **X** which will be the **true length** of lines **A2 and A3.**

On the side elevation it can be seen that the line **E1** is a **true length** and this size can be used in all **4** positions **E1, F4 etc.**

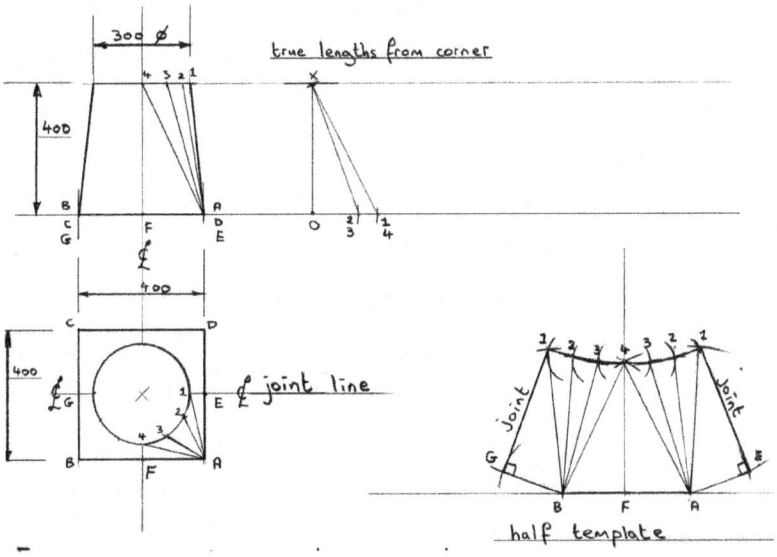

Fig 6-1b Square to round

The half template can now be marked.

A line **AB** is drawn **400** long.

With the dividers or trammel points at the **true length A1**, two intersecting arcs are drawn from points **A and B** to produce point **4.**

With the dividers or trammel points set at the distance 1-2 on the plan view (which is the same as **2-3 and 3-4**) an arc is drawn on either side from point **4.**

With the dividers or trammel points set at the **true length A2** an arc drawn from points **A and B** to intersect at both points marked **3.**

From these two points the arc length **1-2** is marked again.

With the dividers or trammel points still set at **true length A2** an arc is drawn from points **A and B** to intersect at both points marked **2.**

From these two points the arc length **1-2** is marked again.

With the dividers or trammel points set at **true length A1** an arc is drawn from points **A and B** to intersect at both points marked **1.**

With the dividers set at the distance **AE** on the plan view (which is the same as **BG**) two arcs are drawn from points **A and B.**

With the dividers or trammel points set at **true length E1** (which is found on the side elevation) an arc is drawn from both points marked **1** to intersect at points **E and G.**

Since the angles at G and E are right angles in the plan view, they must be right angles in the finished template.

The points marked **1,2,3 and 4** are joined with a fluid curved line to complete the half template, a mirror image of which will be required for the full transition piece.

6c. Inverted square to round.

The inverted square to round is the same as the last example with the exception being, that the square end is now smaller than the round end.

Example.

Fig 6-1c shows a square of **400 x 400** becomes **500 diameter** at a height of **400,** which is made in two halves.

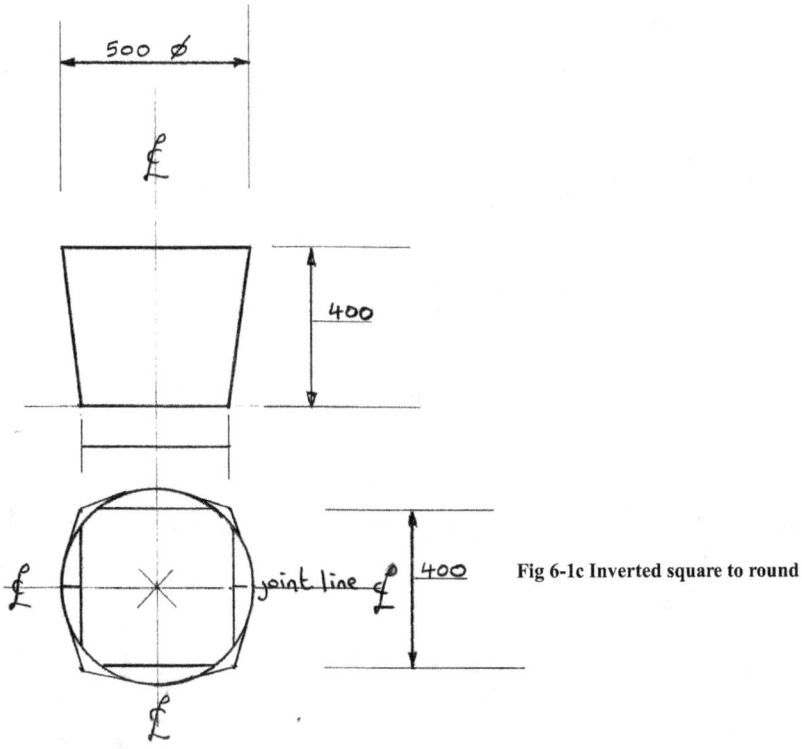

Fig 6-1c Inverted square to round

As before **true lengths** will need to be determined and also, as before it can be seen that the transition piece is symmetrical, therefore only one corner will be used.

Fig 6-2c shows the diameter again divided into **12** equal parts.

A perpendicular **OX** is drawn to the same height as the transition which is **400.**

From the plan view the lengths **A1 and A2** are marked from **O** and drawn back to **X** to produce the **true lengths A1 and A2.**

E1 on the side elevation is a **true length.**

Mark out the half template:

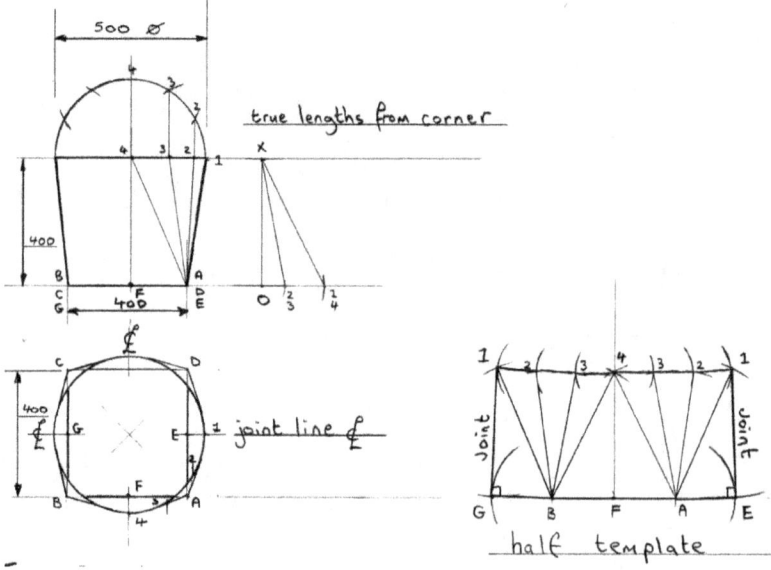

Fig 6-2c Inverted square to round template

The line **AB** is drawn at **400** long.

True length A1 is used as an arc from points **A and B** to intersect at point **4.**

The chord length **1-2** on the plan view is used to describe two arcs from point **4.**

True length A2 is used to describe two arcs from points **A and B** to intersect at points **3.**

The chord length **1-2** is used to describe two arcs from both points marked **3.**

True length A2 is again used to describe arcs from points **A and B** to intersect at points **2.**

The chord length **1-2** is used to describe two arcs from both points marked **2.**

True length A1 is used to describe arcs from points **A and B** to intersect at points **1.**

The length **AE** is used to describe arcs from points **A and B.**

The **true length E1** is used as an arc from both points marked **1** to produce the intersection points at **G and E.**

Since the angles at G and E are right angles on the plan view, they must be right angles in the finished template.

The points **1 to 4** are joined with a fluid curved line to complete the half template. A mirror image of which is required for the full transition piece.

6d. Oblique rectangle to round.

Transitions come in all shapes and sizes but the same rules still apply. Templates for symmetrical shapes can be made mostly with the use of one or two corners however there are exceptions, which occur when the top and bottom centres do not line up with each other.

Example.

Fig 6-1d shows a rectangle to round with the centre of the diameter nearer to one corner of the rectangle. In this case all four corners will need to be dealt with separately.

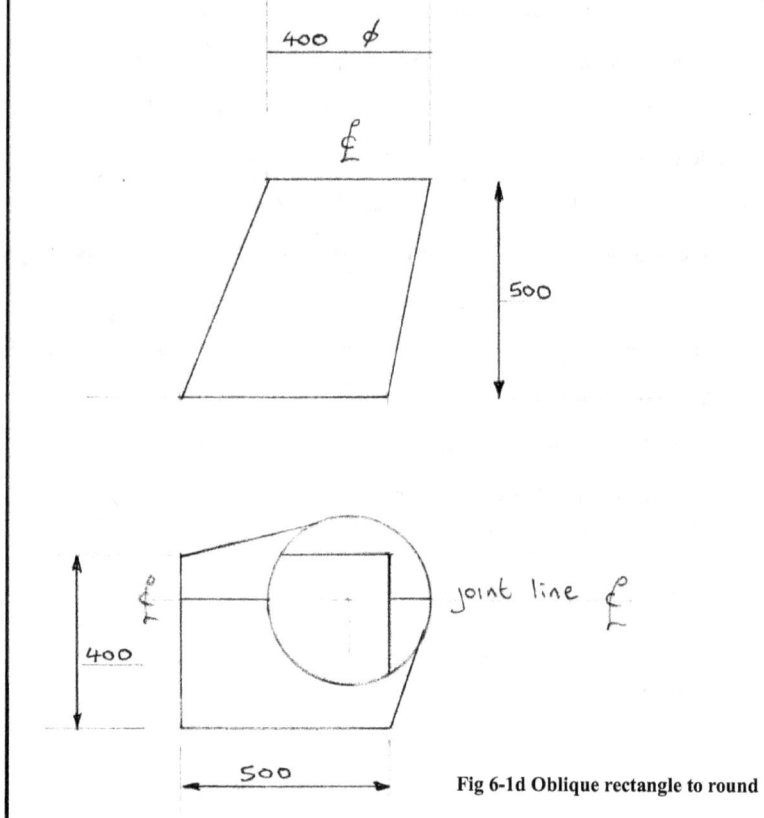

Fig 6-1d Oblique rectangle to round

In **Fig 6-2d** the four corners of the rectangle have been labelled **A,B,C and D** and the

intersection points of the diameter centre lines being **E,F,G and H.**

The diameter has been divided into **12** equal parts numbered **1-12.**

As before perpendiculars will be drawn to find true lengths.

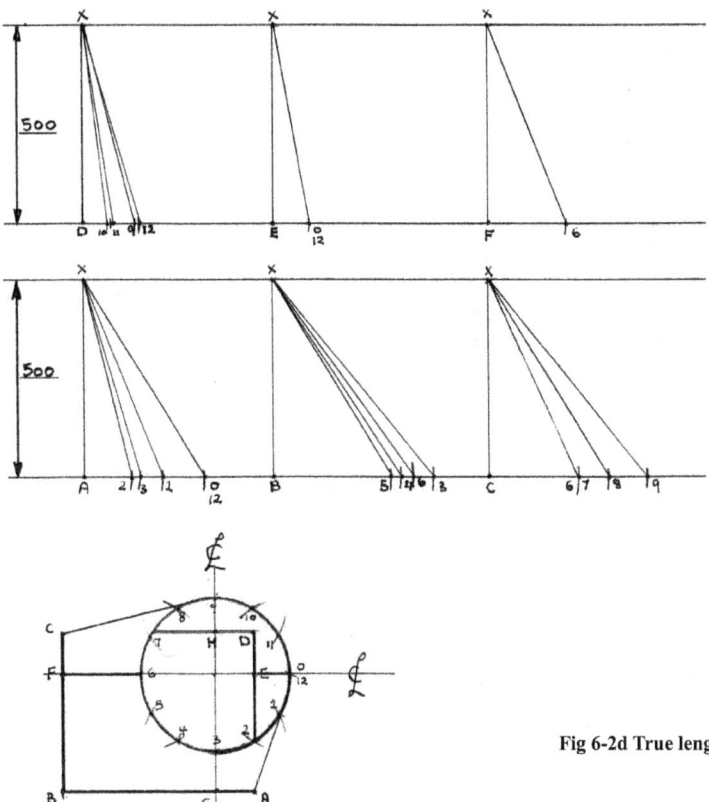

Fig 6-2d True length at corners

Corner A.

A perpendicular is drawn to the height of the transition and marked **AX.**

From the plan view the sizes **A** to **0, 1, 2 and 3** are used as arcs from point **A** and lines are drawn

back to **X** which will be the **true lengths** of **A0, A1, A2 and A3.**

Corner B.

A perpendicular is drawn to the height of the transition and marked **BX.**

From the plan view the sizes **B** to **3, 4, 5 and 6** are used as arcs from point **B** and lines are drawn

back to **X** which will be the **true lengths** of **B3, B4, B5 and B6.**

Corner C.

A perpendicular is drawn to the height of the transition and marked **CX.**

From the plan view the sizes **C** to **6, 7, 8 and 9** are used as arcs from point **C** and lines are drawn

back to **X** which will be the **true lengths** of **C6, C7, C8 and C9.**

Corner D.

A perpendicular is drawn to the height of the transition and marked **DX.**

From the plan view the sizes **D** to **9, 10, 11 and 12** are used as arcs from point **D** and lines are

drawn back to **X** which will be the **true lengths** of **D9, D10, D11 and D12.**

The **true lengths** of the joint lines on the centre line **EF** will also be required for the template.

A perpendicular is drawn to the height of the transition and marked **EX.**

From the plan view the size **E0** is used as an arc from point **E** and a line is drawn back to **X**

which is the **true length** of **E0.**

A perpendicular is drawn to the height of the transition and marked **F6.**

From the plan view the size **F6** is used as an arc from point **F** and a line is drawn back to **X** which is the **true length** of **F6.**

With all true lengths now found the two halves of the template can be drawn.

In **Fig 6-3d** the two halves of the template are drawn next to each other to ensure no mistake is made with the production of the required mirror image (left hand / right hand).

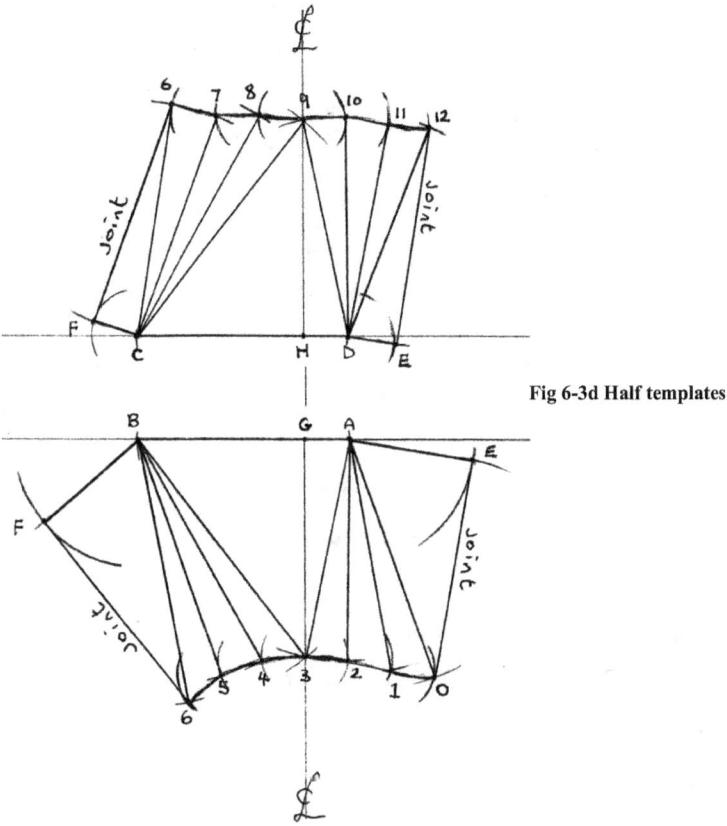

Fig 6-3d Half templates

1st half.

The line **AB** is drawn **500** long.

True length A3 is used to describe an arc from point **A**.

True length B3 is used as an arc from point **B** to intersect and produce point **3**.

Chord length **0-1** on the plan view used to describe an arc on either side of point **3**.

True length A2 is used as an arc from point **A** to intersect at point **2**.

Chord length **0-1** is used to describe an arc from point **2**.

True length A1 is used as an arc from point **A** to intersect at point **1**.

Chord length **0-1** is used to describe an arc from point **1**.

True length A0 is used as an arc from point **A** to intersect at point **0**.

From the plan view the size **AE** is used to describe an arc from point **A**.

The true length E0 is used as an arc from point **0** to intersect at point **E**.

The angle at point E will be a right angle.

True length B4 is used as an arc from point **B** to intersect at point **4**.

The chord length **0-1** is used to describe an arc from point **4**.

True length B5 is used as an arc from point **B** to intersect at point **5**.

Chord length **0-1** is used to describe an arc from point **5**.

True length B6 is used as an arc from point **B** to intersect at point **6**.

From the plan view the size **BF** is used to describe an arc from point **B**.

True length F6 is used as an arc from point **6** to intersect at point **F**.

The angle at point F will be a right angle.

The points **0-6** are joined with a fluid curved line to complete the half template.

2nd half.

A line **CD** is drawn **500** long.

True length C9 is used to describe an arc from point **C**.

True length D9 is used as an arc from point **D** to intersect and produce point **9**.

Chord length **0-1** is used to describe an arc on either side of point **9**.

True length C8 is used as an arc from point **C** to intersect at point **8**.

Chord length **0-1** is used to describe an arc from point **8**.

True length C7 is used as an arc from point **C** to intersect at point **7**.

Chord length **0-1** is used to describe an arc from point **7**.

True length C6 is used from point **C** to intersect at point **6**.

From the plan view the size **CF** is used to describe an arc from point **C**.

True length F6 is used as an arc from point **6** to intersect at point **F**.

The angle at point F will be a right angle.

True length D9 is used as an arc from point **D** to intersect at point **10**.

Chord length **0-1** is used to describe an arc from point **10**.

True length D10 is used as an arc from point **D** to intersect at point **11**.

Chord length **0-1** is used to describe an arc from point **11**.

True length D11 is used as an arc from point **D** to intersect at point **12**.

From the plan view **DE** is used to describe an arc from point **D**,

True length E12 is used as an arc from point **12** to intersect at point **E**.

The angle at point E will be a right angle.

The points **6-12** are joined with a fluid curved line to complete the half template.

CHAPTER 7

Development

A sound knowledge of development is necessary for marking out extra large transition pieces and intersections of pipes to pipes, pipes to hoppers and ducting of any shape into any other shape. The basics learned in this chapter can be applied to most problems the boilermaker should encounter.

7a Over-sized cone sections.

Fig 7-1a shows a very large cone which will need to be made in sections. The size of any particular job will dictate the number of sections required. For this exercise the cone will be produced in four equal parts.

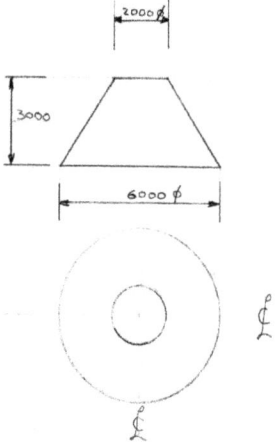

Fig 7-1a Over-sized cone

There are two methods that can be used to produce the quarter cone section, the first is the **true** length cross over method and the second is the **radius without a centre** method, both will be used on this same example.

True length cross over.

Example.

Fig 7-2a shows the information that can be determined by using the information given in **Fig 7-1a.**

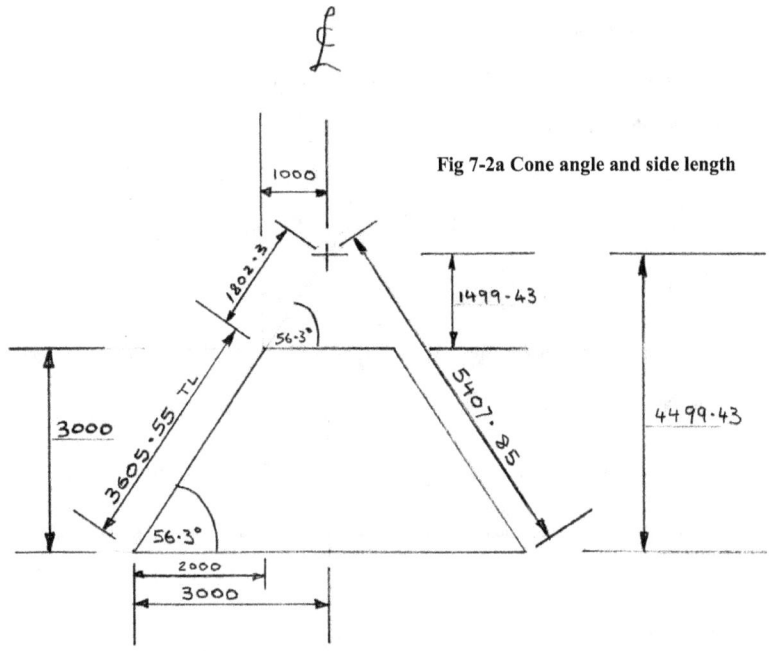

Fig 7-2a Cone angle and side length

Cone angle is determined by..... $\tan = \dfrac{opp}{hyp}$ .... $\tan = \dfrac{3000}{2000}$ .... **56.3°.**

True length of side is determined by .... $a^2 + b^2 = c^2$........ $3000^2 + 2000^2 = 13000000$ ....

$\sqrt{13000000} = \mathbf{3605.55.}$

Small radius of cone is determined by..... $hyp = \dfrac{adj}{\cos}$ ..... $hyp = \dfrac{1000}{\cos 56.3}$ ...... hyp = **1802.3.**

Large radius of cone is determined by........ **3605.55 + 1802.3 = 5407.85.**

Height from top of cone to apex is determined by..... $c^2 - a^2 = b^2$ ....

$3248285.29 - 1000000 = 2248285.29$.............. $\sqrt{2248285.29} = \mathbf{1499.43.}$

Height from cone base to apex is determined by.... 3000 + 1499.43 = **4499.43.**

Information contained in Fig 7-2a can now be used to determine lengths and chords used in the section development.

The finished cone has a base diameter of **6000.**

The circumference of which is..... $\mathbf{\pi D}$........ π x 6000 = **18849.55.**

The length of arc for **90°**, or, a quarter section is ...... $\dfrac{18849.55}{4}$ = **4712.38**

The large radius of the cone is **5407.85** which when multiplied by **2** gives a diameter of **10815.7.**

The circumference of which is π x 10815.7 = **33978.52.**

The arc length for **1°** .... $\dfrac{33978.52}{360}$ = **93.384.**

With this information the angle of arc needed for the quarter cone section can be determined to be.......

$\dfrac{4712.38}{94.384}$ = **49.92°.**

Fig 7-3a shows the quarter cone section bounded by the angle of **94.92°**.

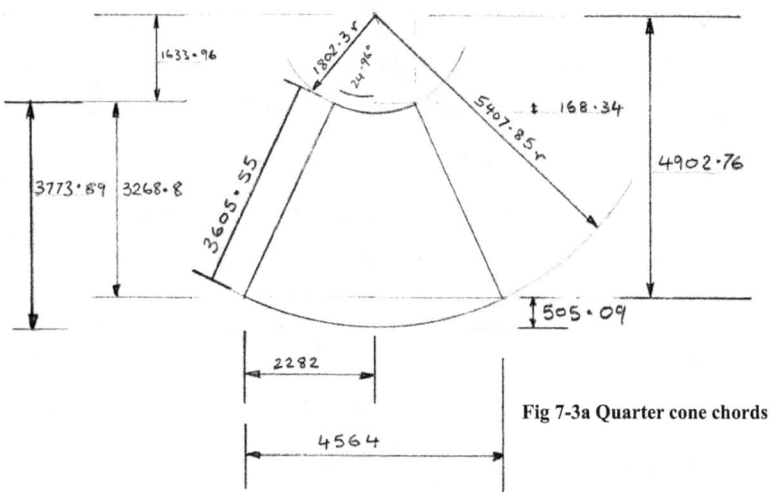

Fig 7-3a Quarter cone chords

The chords and heights are determined as follows:

$$\frac{49.92}{2} = 24.96°.$$

opp = sin x hyp........ opp = sin 24.96 x 5407.85 = **2282**.

This is half of the bottom chord, which when multiplied by **2** gives the full chord of **4564**.

opp = sin 24.96 x 1802.3 = **760.54.**

This is half of the top chord, which when multiplied by **2** gives the full chord of **1521.**

height from top chord to apex is determined by.... **adj = cos x hyp**

.... adj = cos 24.96 x 1802.3 .... = **1633.96.**

height from bottom chord to top chord is determined by......

adj = cos 24.96 x 3605.55 .... = **3268.8.**

height from bottom chord to apex is determined by 1633.96 + 3268.8 = **4902.76.**

height between bottom chord and radius is determined by 5407.85 minus 4902.76 = **505.09.**

height between top chord and radius is determined by 1802.3 minus 1633.96 = **168.34.**

Fig 7-4a shows the quarter cone development in the centre in small scale.

Both the top and bottom radius have been divided into 8 equal spaces.

Any number of spaces can be used as long as the number is **the same for both ends**.

The drawing on the left of the page shows the segment between **4 and 5** on the large radius and

X and Y on the small radius.

The angle of arc of the section is determined by dividing the angle of the full development by

the number of divisions used .... $\frac{49.92}{8}$ = **6.24°.**

Half of this angle together with the **true length** of the cone side is used to determine the height

between the chords .... cos 3.12 x 3605.55 = **3600.**

The length of top chord is determined by.... opp = sin x hyp.......

opp = sin 3.12 x 1802.3 = **98.09.**

This is half of the top chord, which when multiplied by **2** gives the full chord of **196.188.**

The length of the bottom chord is determined by opp = sin 3.12 x 5407.85 = **294.33.**

This is half of the top chord, which when multiplied by **2** gives the full chord of **588.67.**

Refer to the drawing on the right of **Fig 7-4a.**

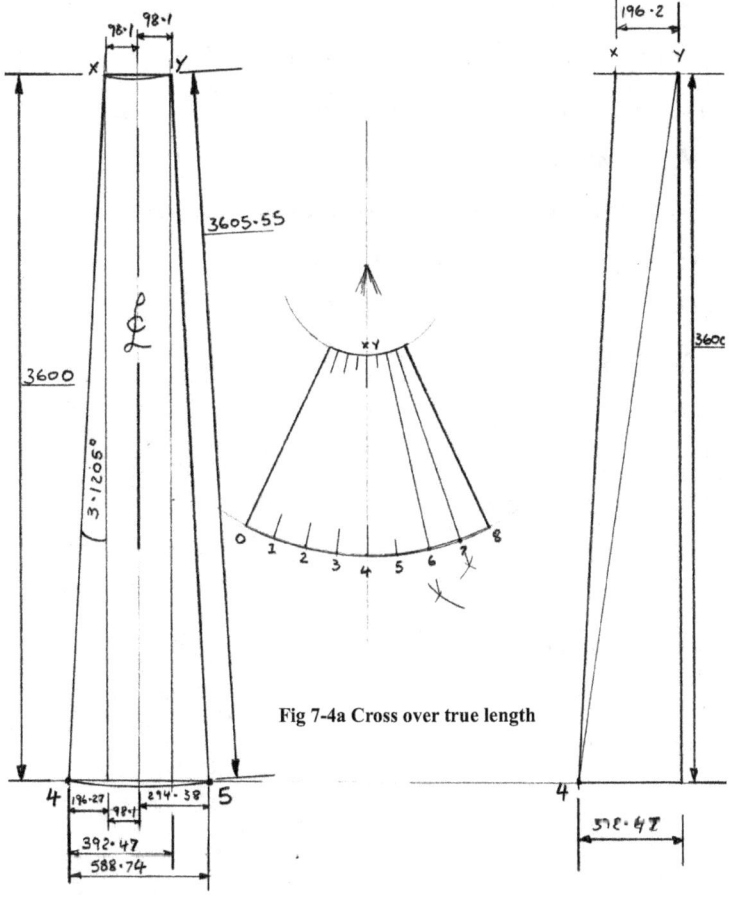

Fig 7-4a Cross over true length

The final piece of information needed for the **true length cross over** method is the **true length** between **4** on the bottom chord and **Y** on the top chord, which is the same as **5X.**

The first thing to do is determine the length of the base..............

opp = sin x hyp .... opp = sin 3.1205° x 3605.55 = **196.27.**

base = 196.27 + top chord .... 196.27 + 196.2 = 392.47.

The length **4Y** is determined by.... $c^2 = a^2 + b^2$ .... $392.47^2 + 3600^2 = 13114032.7$..............

$\sqrt{13114032.7} = $ **3621.33.**

It is now a simple matter of transferring known sizes to the development.

With reference to the diagram in the centre of **Fig 7-4a**................................

A perpendicular is drawn to the length of **4X** (3605.55).

From **X** the chord **XY** (196.2) is used to describe an arc from **X.**

From **4** the length **4Y** (3621.33) is used as an arc to intersect at **Y.**

From **4** the chord **4, 5** (588.74) is used to describe an arc from **4.**

From **X** the length **4Y** (3621.33) is used as an arc to intersect at **5.**

This development is the segment of the cone between **4 and 5 (bottom) and X and Y (top).**

The process is repeated on one side until number **8** is reached and on the other side until **0** is reached to complete the quarter cone.

A fluid line joins all points at the top and all points at the bottom to finish the development, which will need to be made four times for the full cone.

Radius without a centre.

Example.

With reference to **Fig 7-3a** the sizes for the top and bottom chords are used to mark the top and

bottom radius.

In **Fig 7-5a** the bottom chord with the length **4564** is marked **AB** with the height **505** marked as a

perpendicular at the centre to produce point **C.**

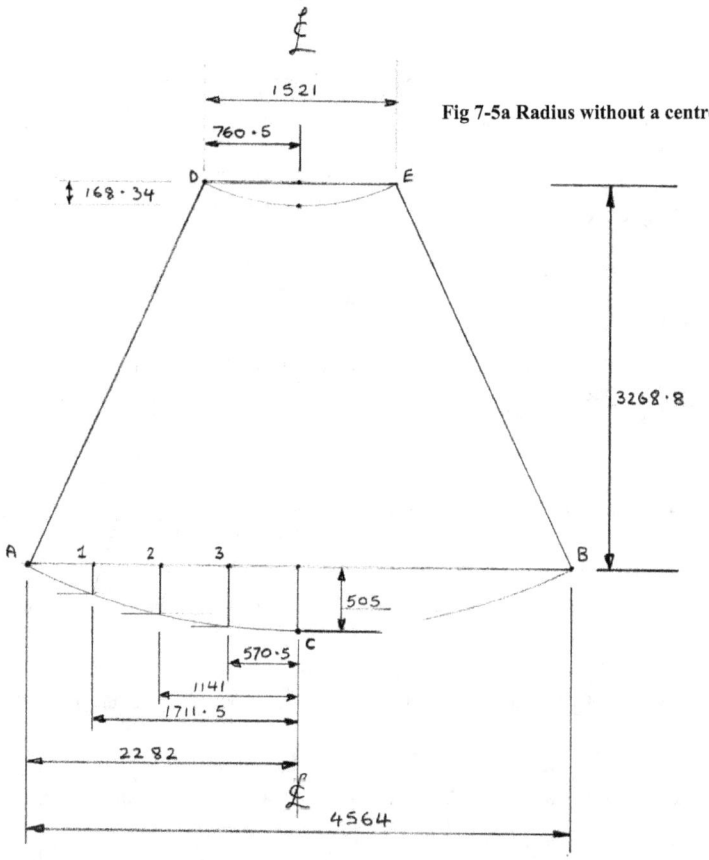

Fig 7-5a Radius without a centre

Since the curve will be symmetrical only one half of the curve will be developed and then reproduced on the other side.

The length of half of the chord (**2282**) is divided into any number of equal parts. For this exercise **4** will be used.

$$\frac{2282}{4} = 570.5.$$

The base length 2282 with the hypotenuse length 5407.85 in a right angled triangle gives a third side length of 4902.7, which when subtracted from the hypotenuse gives a difference of 505.

It is this difference that will be used to plot the curve of the development.

2282 minus 570.5 = 1711.5

1711.5 minus 570.5 = 1141

1141 minus 570.5 = 570.5

With these sizes used as the base length of a right angled triangle with the hypotenuse length of 5407.85 the third side of each triangle is subtracted from the hypotenuse to give the sizes needed to plot the curve.

The base length 1711.5 gives a third side of 5129.8. The difference between this and 5407.85 is **278.**

The base length 1141 gives a third side of 5286. The difference between this and 5407.85 is **121.85.**

The base length of 570.5 gives a third side of 5377.6. The difference between this and 5407.85 is **30.25.**

In order for these sizes to be marked from the chord AB they must be subtracted from 505.

In **Fig 7-5a** the height at point **3** is 505 minus 30.25 = **474.75.**

The height at point **2** is 505 minus 121.85 = **383.15.**

The height at point **1** is 505 minus 278 = **227.**

The same points are plotted on the other half of the chord from **C** to **B** and all points are joined with a fluid curve to complete the bottom radius.

A perpendicular line is drawn from the centre of **AB** to a height of **3268.8** (fig 7-3a) and line parallel to **AB** is drawn.

Half of the chord length **1521** is marked on each side of the centre line to produce **D** and **E.**

The size of **168.34** is marked on the centre line, and half of the chord is divided into an appropriate number of parts so that the top curve can be plotted in exactly the same way as the bottom curve, this time using the value **1802.3** (fig 7-3a) for the hypotenuse.

7b Lobster back bends.

Lobster back bends are used where it is not practical to use regular mass produced bends. Most industrial ducting is constructed using lobster back bends or segmented bends as they are sometimes called. In this exercise the bend will be made in **3** segments but any number can be used to produce any desired angle.

The segments can be cut from an existing pipe by making a template and wrapping it around the pipe to mark the cut line, or can be rolled from a plate with the development marked directly onto the plate.

Example.

Fig 7-1b shows a **90°** bend is **400** diameter and made up of **3** segments. Each segment will be

30°.

The centres of the bend will be **600.**

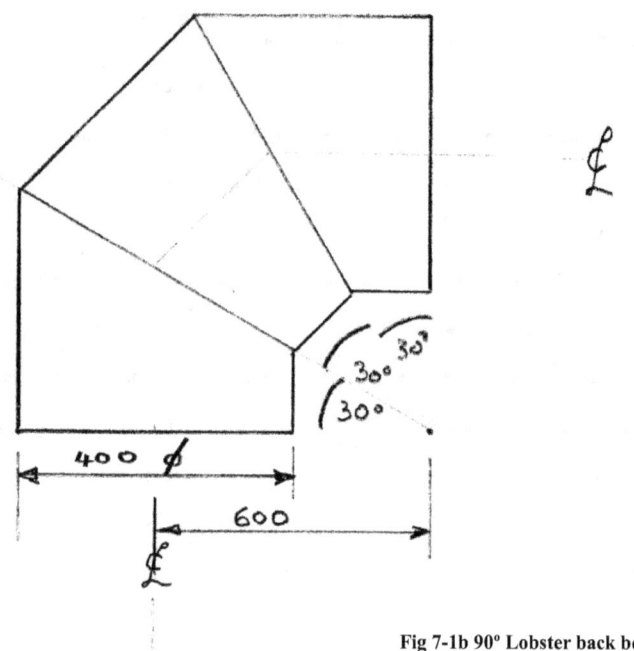

Fig 7-1b 90° Lobster back bend

Fig 7-2b shows the circumference divided into **12** equal parts with these divisions marked throughout the entire bend and are numbered to avoid confusion.

A centre line has been added to the centre section to act as a base line to transfer sizes to the template.

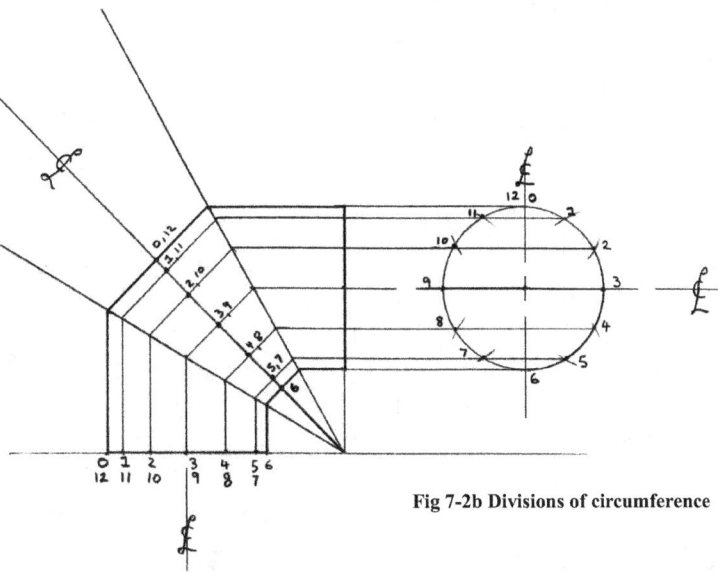

Fig 7-2b Divisions of circumference

The two outside sections are identical therefore one template will serve for both sections.

One template will be drawn for the outside sections and one template will be drawn for the centre section.

The circumference must be determined for the length of the template. **It must be remembered that the outside diameter is used for wrap around, and mean diameter is used for plates to be rolled.**

Circumference = πD............. π x 400 = **1256.6.** This figure is the length of each template.

Divided by the number of divisions will give the distance between each line of the template.

$\frac{1256.6}{12}$ = **104.7. Fig 7-3b** shows both templates with the divisions @ 104.7 marked as

perpendiculars on the full length of 1256.6.

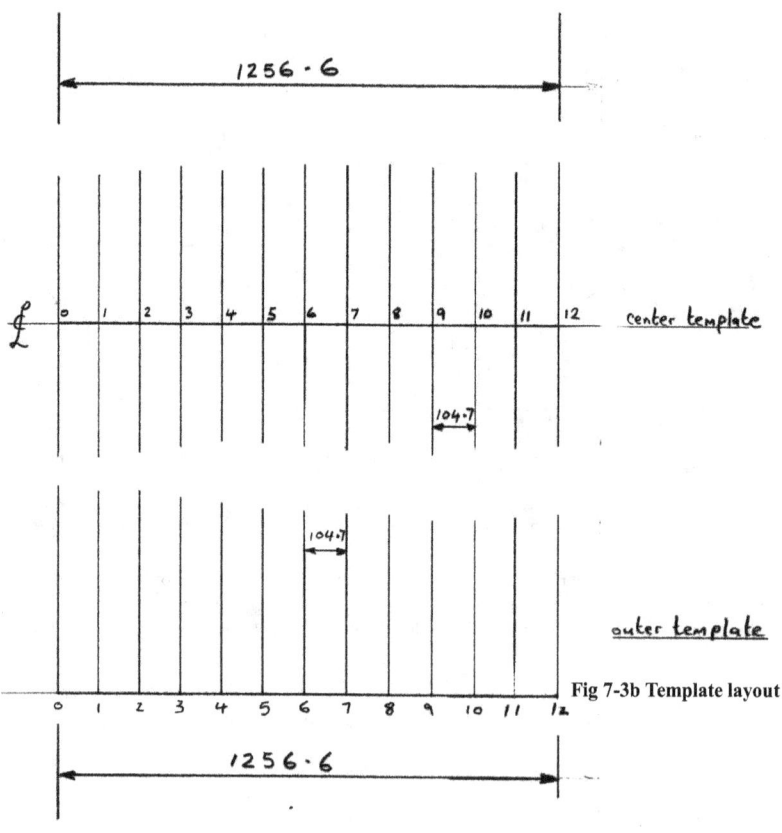

Fig 7-3b Template layout

The outside template needs only one side of the base line to be marked, whereas the centre template must be marked on both sides.

It must be remembered that the point where the template starts and finishes will be the joint line.

For example, if this bend must have all joint welds at the narrowest part of each section, each template would start and finish at point **6.** If the welds were required to be at the widest section, the templates would start and finish at **0, 12. The weld lines are usually a combination of both.** **Fig 7-4b** shows both templates starting and finishing at **0, 12.**

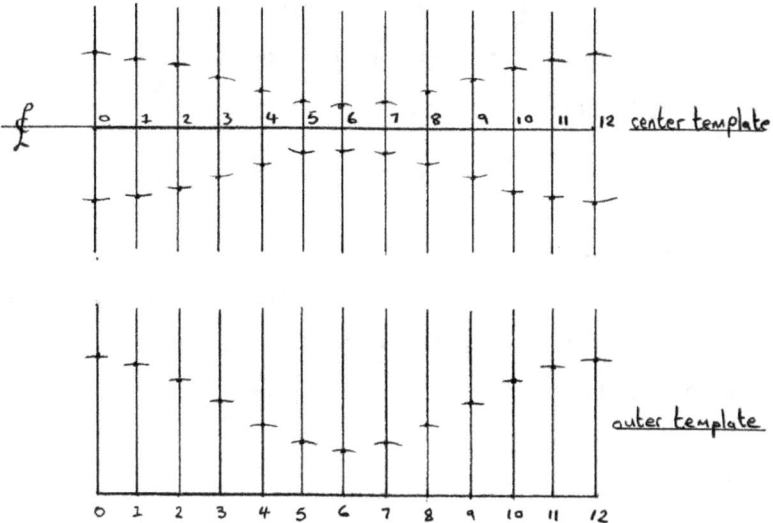

Fig 7-4b Sizes transferred

Outer section.

All sizes are transferred from **fig 7-2b** to the template layout **fig 7-4b.**

With the dividers set at the distance of point 0 to the intersection line at 30°, transfer this size to the outer template at point **0.**

Repeat the process with all other points until point **12** is reached.

A fluid curved line is drawn to complete the template.

Centre section.

With the dividers set at the distance 0 on the centre line to either intersection line at 30°, transfer the size to both sides of the base line (centre) on the centre template.

Repeat the process with all other points until point **12** is reached.

A fluid curved line is drawn to complete the template.

7c Intersections of pipes.

Making templates for intersections of pipes follows the same principles used in lobster backs.

The templates used are generally the wrap around type therefore the circumference is calculated using the outside diameter of the pipe.

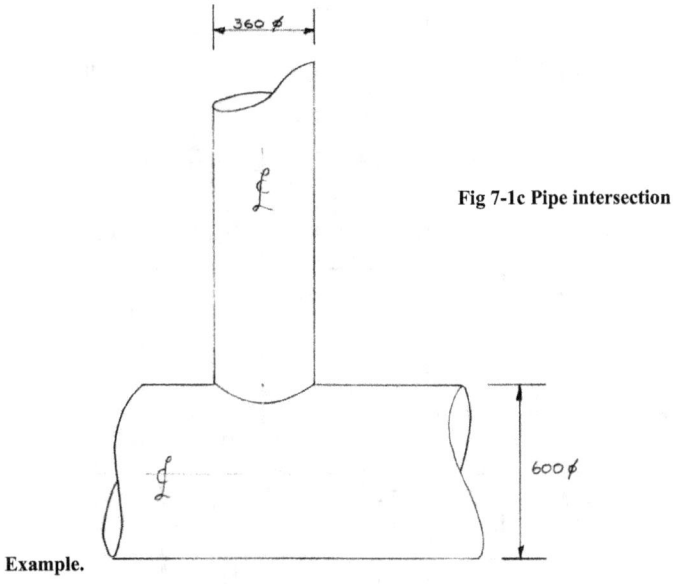

360 ⌀

600 ⌀

Fig 7-1c Pipe intersection

Example.

Fig 7-1c shows a **360** diameter pipe intersecting with a **600** diameter pipe at **90°.**

The small pipe will need to be divided into a number of equal divisions. The size of the pipe will

dictate the number of divisions, more for large pipes and less for smaller pipes.

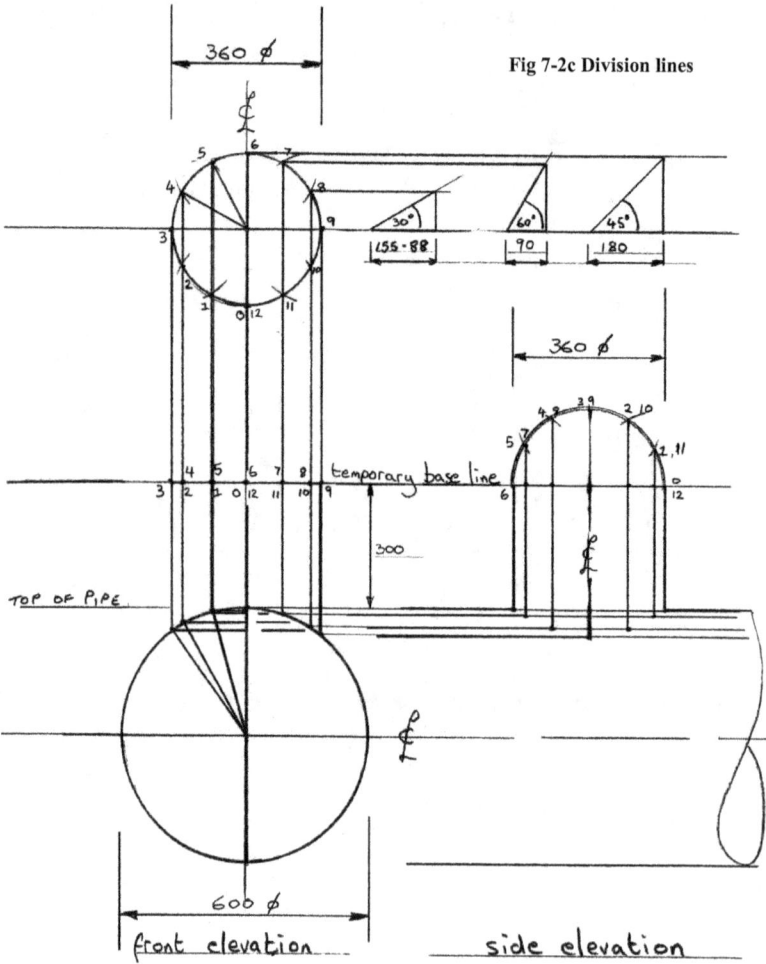

Fig 7-2c Division lines

360 ∅

155·88   30°
90   60°
180   45°

360 ∅

temporary base line

300

TOP OF PIPE

front elevation   side elevation

600 ∅

A temporary base line at a distance of **300** from the large pipe is used for the smaller template.

Fig 7-2c shows the smaller pipe divided into **12** equal parts, which will give divisions of **30°**.

Two views are shown, both the front and side elevation with the divisions on the small pipe to show the intersection points.

The diameter of the smaller pipe gives a circumference of ...... π x 360 = **1130,97**.

$\dfrac{1130.97}{12}$ = **94.24**, which is the distance between the template lines.

Fig 7-3c shows the template layout for the small pipe, with the **12** division lines drawn as perpendiculars.

As before all sizes are transferred from **fig 7-2c.**

With the dividers set at the distance **0** on the temporary base line to the intersection point with the 600 diameter pipe, transfer this size to the perpendicular at point **0** in **fig 7-3c.**

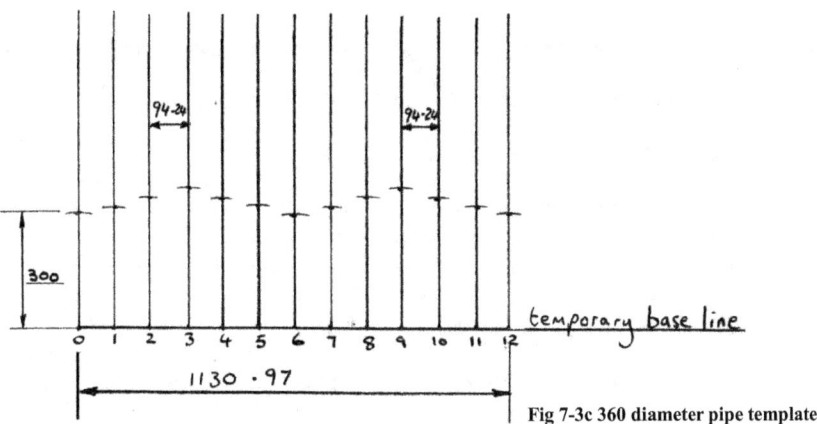

Fig 7-3c 360 diameter pipe template

With the dividers set at the distance **1** to the intersection point with the 600 diameter pipe, transfer this size to the perpendicular at point **1** in **fig 7-3c.**

Continue in this manner until point **12** has been reached.

A fluid curved line is drawn to complete the template.

Since the smaller pipe has a diameter of 360, a template **420 wide** will suffice for the template of the hole in the 600 diameter pipe.

The diameter of the larger pipe gives a circumference of .... π x 600 = **1884.95**

Division lines are only necessary at the centre of large pipe development and must coincide with the division lines on the small pipe, as the side view to the right of the main drawing shows.

The method of determining the position of these divisions on the large pipe is shown in **3** stages at the top of the drawing, namely, **30°, 60° and 90°. (fig 7-2c).**

cos = adj x hyp.

An angle of 30° with a hypotenuse of 180 (radius of the small pipe) gives a base of **155.88**

An angle of 60° with a hypotenuse of 180 gives a base of **90.**

An angle of 90° will be the same as the radius, i.e. **180.**

These sizes are now used with the radius of the large pipe to produce angles.

$$\sin = \frac{opp}{hyp}$$

Base **155.88** with a hypotenuse of **300** (radius of large pipe) gives an angle of **31.3°.**

Base **90** with a hypotenuse of **300** gives an angle of **17.45°.**

Base **180** with a hypotenuse of **300** gives an angle of **36.86°.**

These angles are now used to find arc lengths on the large pipe.

The circumference of the pipe is 1884.95.

$$\frac{1884.95}{360} = \textbf{5.23.}\ \text{This is the length of one degree of arc.}$$

31.3 x 5.23 = 163.69.

17.45 x 5.23 = 91.26.

36.86 x 5.23 = 192.77.

Since the development is symmetrical, these sizes are used on either side of the template centre

line as division lines to plot the shape of the hole.

Fig 7-4c shows the layout for the template with the division lines drawn in.

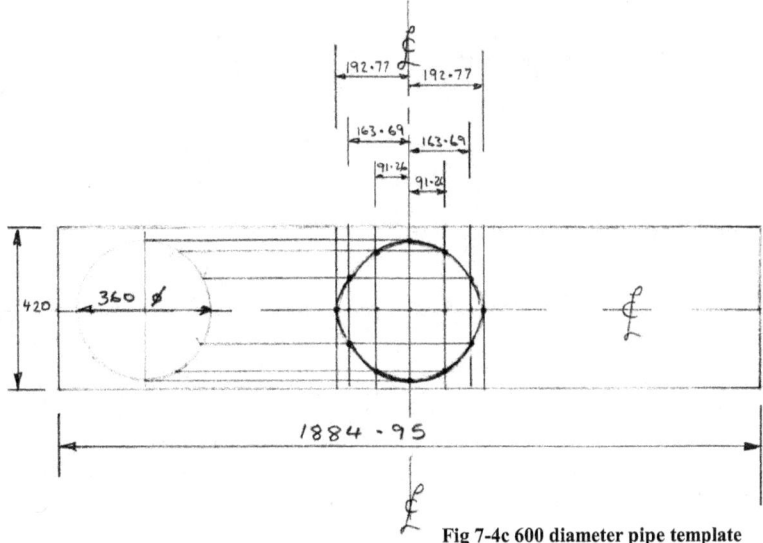

Fig 7-4c 600 diameter pipe template

On the left of the template a circle **360** diameter has been drawn with **12** divisions to intersect

with the lines drawn at the centre.

A fluid curved line is drawn through the intersection points to complete the template.

Inclined pipe intersections.

Example.

Fig 7-5c shows a pipe **400** diameter intersecting with a **610** diameter pipe at **60°.**

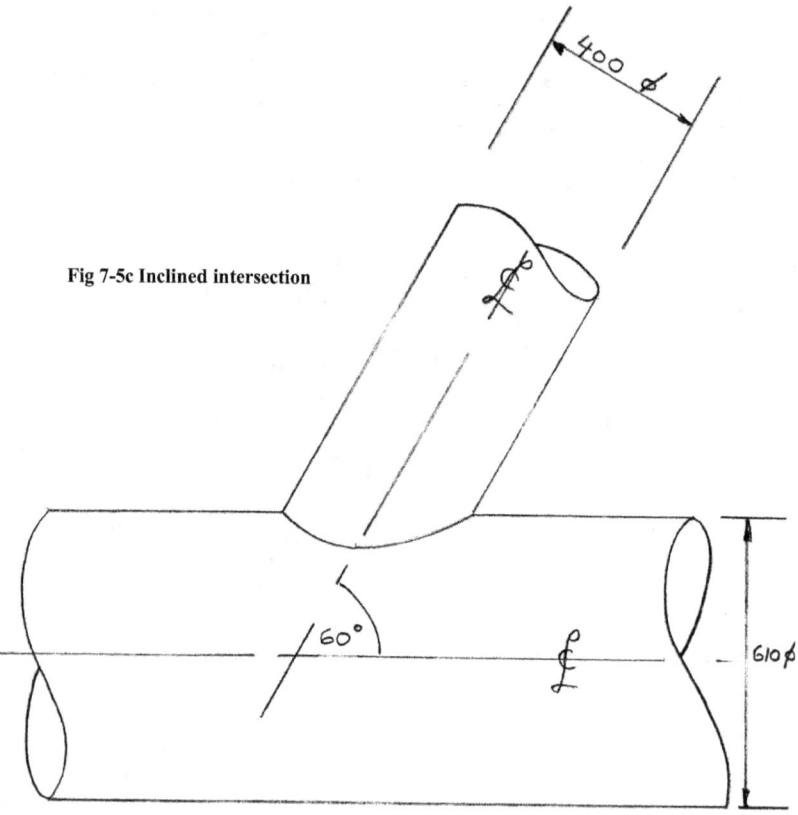

Fig 7-5c Inclined intersection

Again a temporary base line will be used and the smaller pipe will be divided into **12** equal parts.

Fig 7-6c shows both the front and side elevations as in the last exercise.

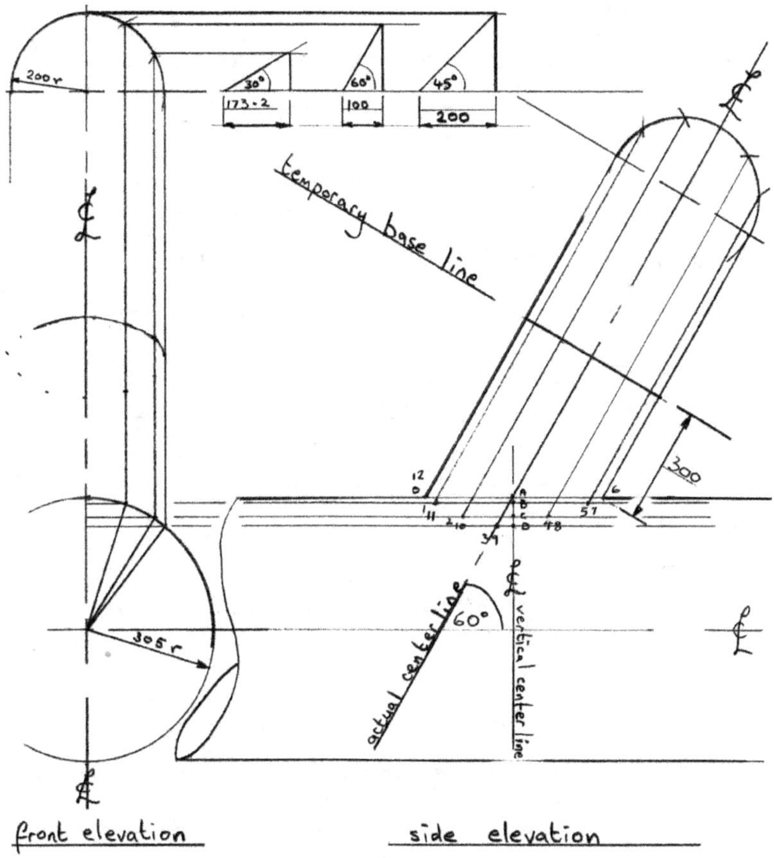

front elevation   side elevation

Fig 7-6c Intersection points

The length of the development of the smaller pipe will be its circumference; π x 400 = **1256.6.**

The distance between the template lines of the smaller pipe is.....

(π x 400 divided by 12) = **104.7**

Fig 7-7c shows the template layout for the small pipe with the **12** division lines marked as perpendiculars and the heights transferred from the temporary base line.

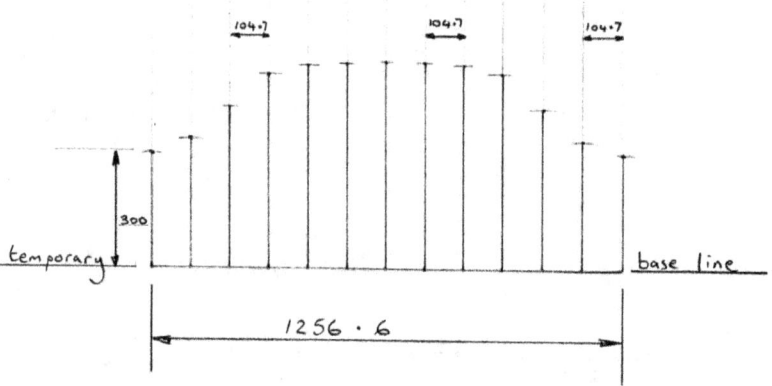

Fig 7-7c 400 Diameter pipe template

All sizes are transferred from **fig 7-6c** in the same manner as the last exercise.

The sizes will be transferred from between the temporary base line in the side elevation and the intersection points with the larger pipe.

The length of the template for the larger pipe has a length of; π x 610 = **1916.3.**

A centre line is marked as a base line as in the previous exercise.

3 lines are then marked on either side with the distances from the centre line being calculated as before.

First find the angle then use it to determine the arc length.

$$\frac{1916.3}{360} = 5.323 = \text{arc length of } 1°. \quad (\sin = \frac{opp}{hyp})$$

The furthest away from the centre line will be at a distance of (**200** which is the radius of the small pipe divided by **305** which is the radius of the large pipe, to give **40.97° x 5.323) 218.**

The middle line is at a distance of $\frac{173.2}{305}$ = **34.6° x 5.323 = 184.17.**

The closest line is at a distance of $\frac{100}{305}$ = **19.139° x 5.323 = 101.87.**

Fig 7-8c shows at the top, the template with these lines marked and also the central vertical line from the side elevation in **fig 7-6c,** which will be used as a base line to mark from. For clarity the starting points on the vertical base line have been marked **A, B, C, and D**, and the intersection points have been marked **0-12.** It can be seen that the only place where the two centre lines coincide is at the top of the large pipe, as the smaller pipe is projected through the large pipe the intersection points get closer to the vertical centre line on one side and further away on the other.

These sizes are transferred to the development from the vertical centre line, with 4-8 being below the vertical centre line and all other points above.

Fig 7-8c shows at the bottom, the finished development of the hole.

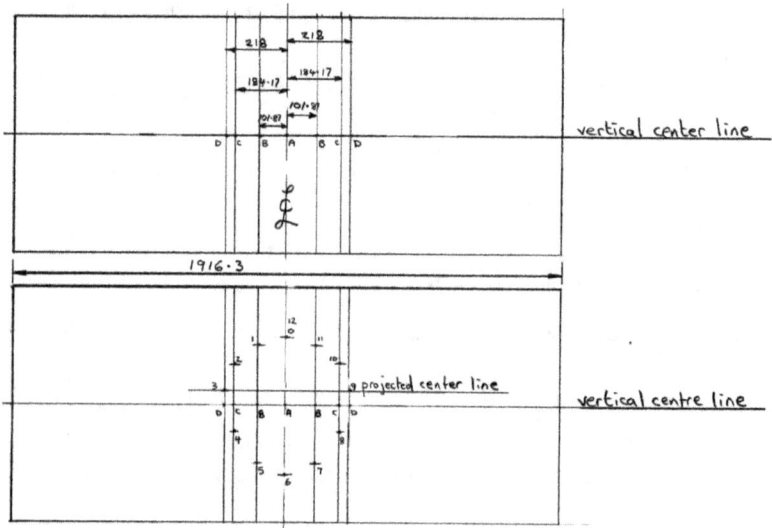

Fig 7-8c 610 Diameter pipe template

7d Intersections of cones.

Intersections of cones can be quite tricky because of the many lines that need to be drawn to find the intersection points. But if enough care is taken to label each line clearly with either numbers or letters, any intersection should be within the scope of the average boilermaker.

The easiest intersection of a cone would be a pipe intersecting vertically aligned centrally with the cone, which would of course produce a circular intersection.

The easiest intersection requiring a template would be a pipe intersecting horizontally with the cone.

Example.

Fig 7-1d shows a large pipe intersecting horizontally with a cone, for this exercise sizes are not important.

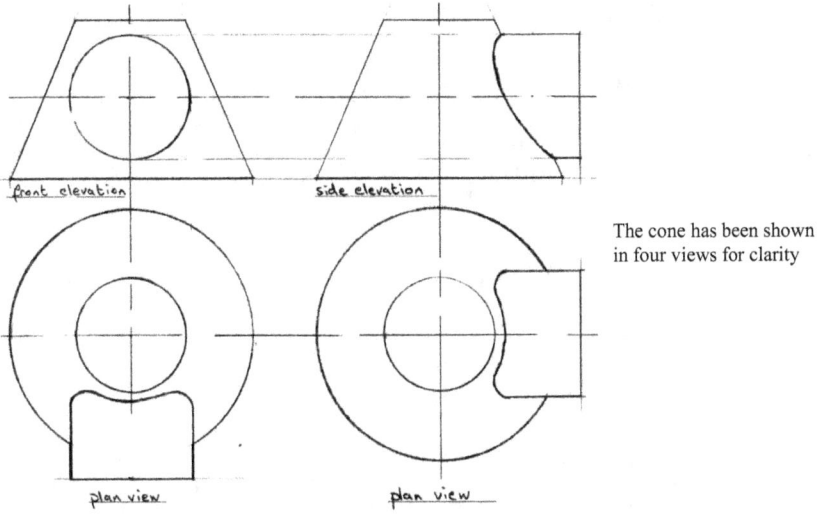

The cone has been shown in four views for clarity

Fig 7-1d Horizontal intersection

A working drawing will need to be produced from the information provided in the workshop drawing. All vertical lines on the cone **must** be drawn to the apex therefore on the working drawing the apex will be shown.

Cone intersections require interactive drawings, meaning more than one view is required to plot the intersection points and each view needs the others for information, as shown in **fig 7-2d.**

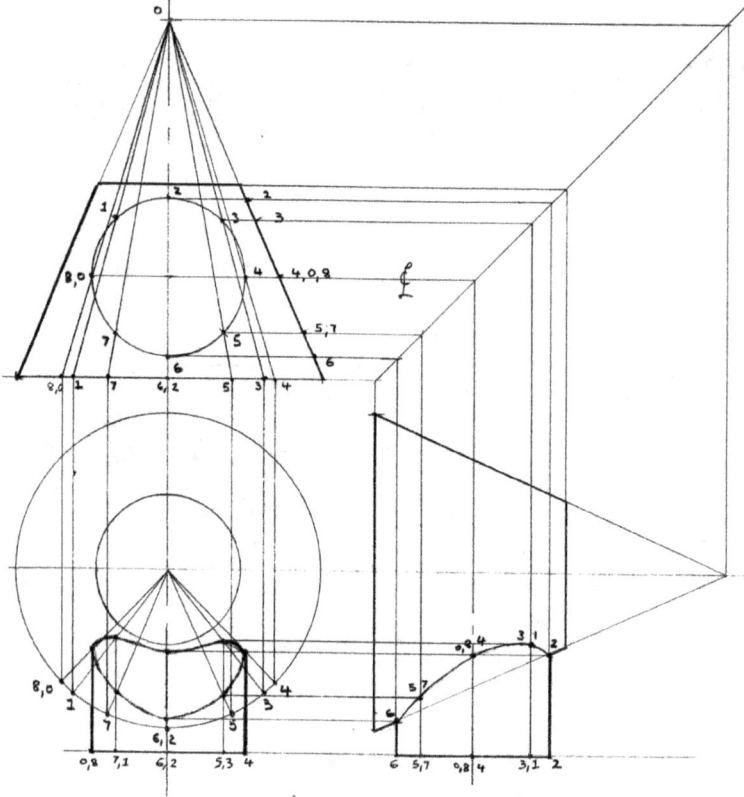

Fig 7-2d Intersection points

The front elevation is at the top with the side elevation to the right and the plan view in the middle. In this configuration all intersecting lines can be drawn and used for transfer to the template.

A **45°** line is used to transfer lines between the front and side elevations.

The circumference of the pipe has been divided into **8** equal parts, (any number can be used) these points have been numbered and drawn back to the apex on the front elevation and also projected to the base of the cone, where they have been numbered to avoid confusion.

From the base of the cone these points are now projected vertically down to the cone circumference in the plan view.

Again they are numbered to avoid confusion and drawn back to the plan view apex just as in the front elevation.

Points **1-8** on the **pipe circumference** are projected vertically down to the end of the **pipe** in the plan view and are numbered to avoid confusion.

Vertical lines are then drawn from these points to intersect corresponding lines drawn from the circumference to the centre.

It will be seen that points **2 and 6** can only be found by transferring them horizontally from the side elevation.

When all **8** points have been plotted on the plan view they can be transferred horizontally to the side elevation to intersect with the corresponding points transferred from the front elevation.

With all intersection points marked in their correct position, the templates can now be made.

Pipe template.

The diameter of the pipe multiplied by π will give its circumference, which will be the length of the template. The circumference divided by **8** will give the distance between the perpendiculars on the template.

All sizes will be transferred directly from the side elevation in **fig 7-2d.**

Starting at **0,** which is on the centre line, the distance between the end of the pipe and each

intersection point, is transferred directly onto the template as in **fig 7-3d.** The pipe template is at

the bottom and the cone template is at the top.

Cone template.

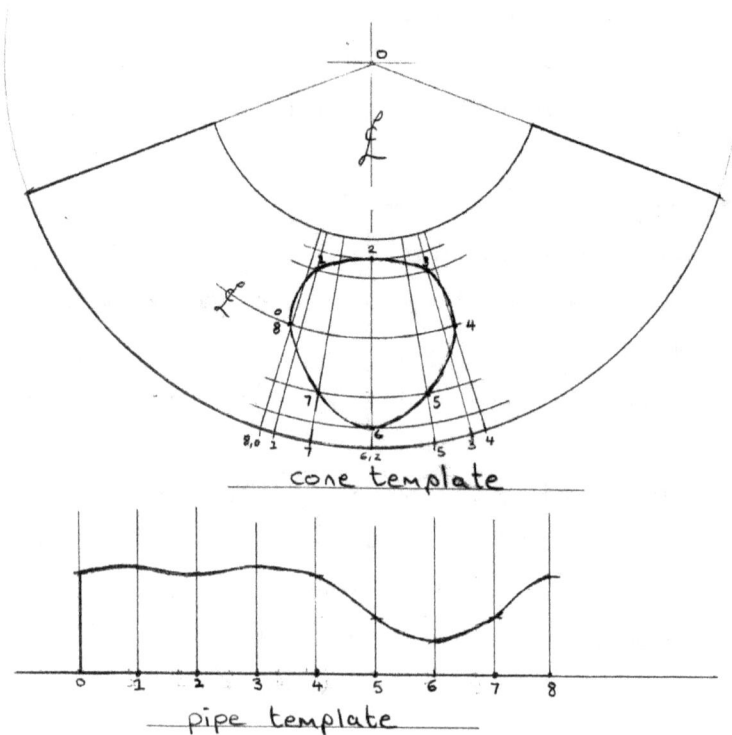

Fig 7-3d Templates

At the top of **fig 7-3d** the cone is marked out as in chapter 5.

A centre line is drawn and at the point where it intersects the circumference, the starting point for **2 and 6** are found.

All sizes are taken from **fig 7-2d.**

From this point with the dividers or trammel points set at the chord distance **2 - 5** in the plan view, points **5 and 7** are marked.

With the dividers or trammel points set at the chord distance **3 - 5,** point **3** is marked from point **5,** and point **1** is marked from point **7.**

With the dividers set at chord distance **3 - 4,** point **4** is marked from point **3,** and point **0 / 8** is marked from point **1.**

A line is drawn from each point back to the apex.

From the apex marked **0** on the front elevation the distance of each point on the **right hand sloping side** of the cone is marked as an arc on the template from the apex, and the relevant intersection points are plotted.

A fluid curved line drawn between all points on both templates completes the exercise.

Inclined intersection.

A pipe intersecting with a cone at an angle is more involved than a straight intersection. If each intersection point is dealt with separately and care is taken to label all lines clearly to avoid confusion, the principles applied in the next exercise can be applied to the majority of cone intersections the boilermaker may encounter.

Example.

Fig 7-4d shows a pipe intersecting with a cone at 45° to the vertical. As before sizes are not necessary as this is a lesson in geometry.

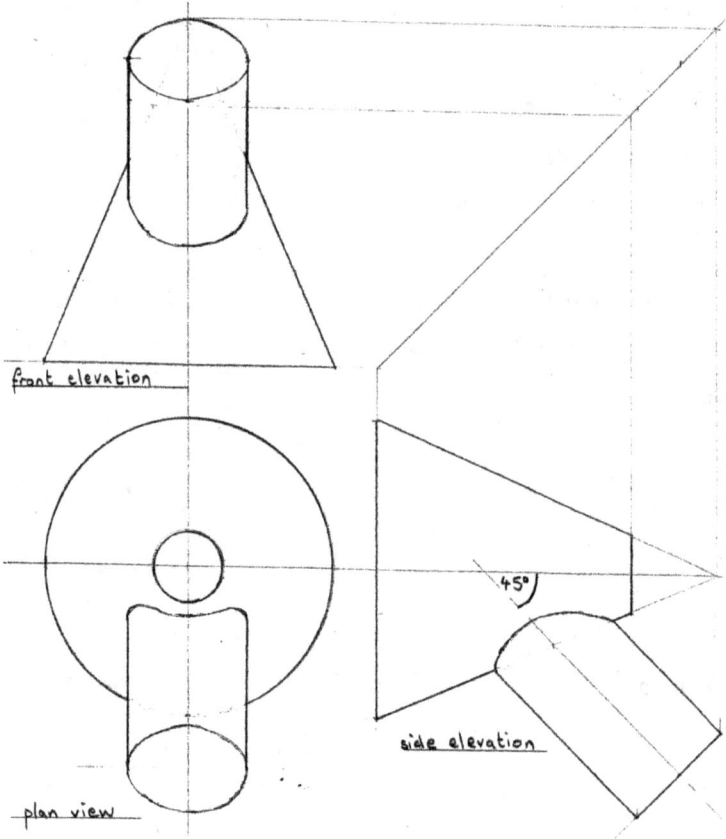

Front elevation

plan view

side elevation

45°

Fig 7-4d Inclined intersection

Fig 7-5d shows the cone circumference divided into **12** equal parts (any number can be used).

These points are projected vertically and horizontally to the base of the cone in the other two

views, the side elevation requires only points **3 - 6.**

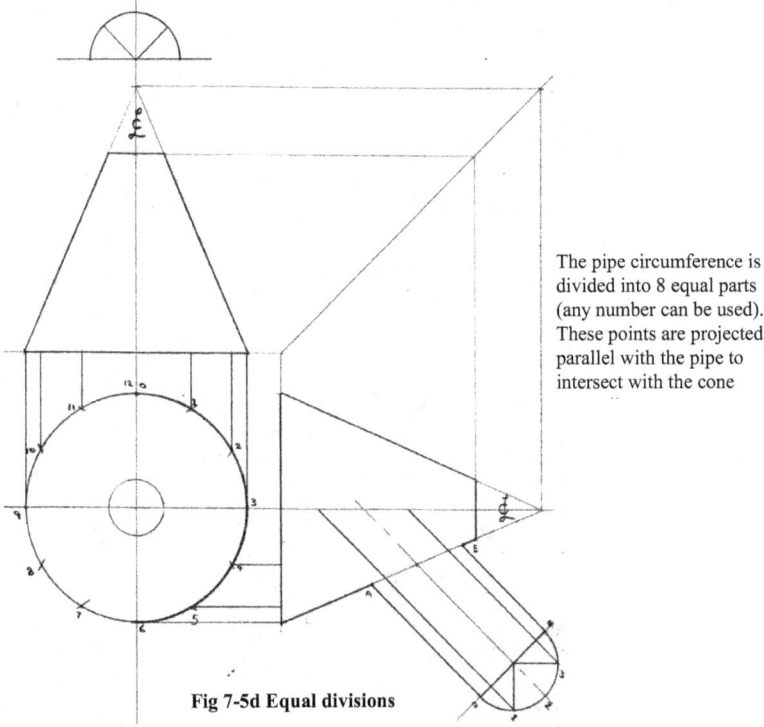

The pipe circumference is divided into 8 equal parts (any number can be used). These points are projected parallel with the pipe to intersect with the cone

Fig 7-5d Equal divisions

Since the figure is symmetrical the half-section of the pipe will use numbers **0 - 4** only.

The intersection points of the lines numbered **1** to **3** will be labelled **B C and D,**

It will be seen that points **A and E** are on the outside (full diameter) of the cone at the

intersection points of lines **0 and 4** which is the full diameter of the pipe.

In **fig 7-6d** the points at the base of the cone have been drawn back to the apex in the front and

side elevations.

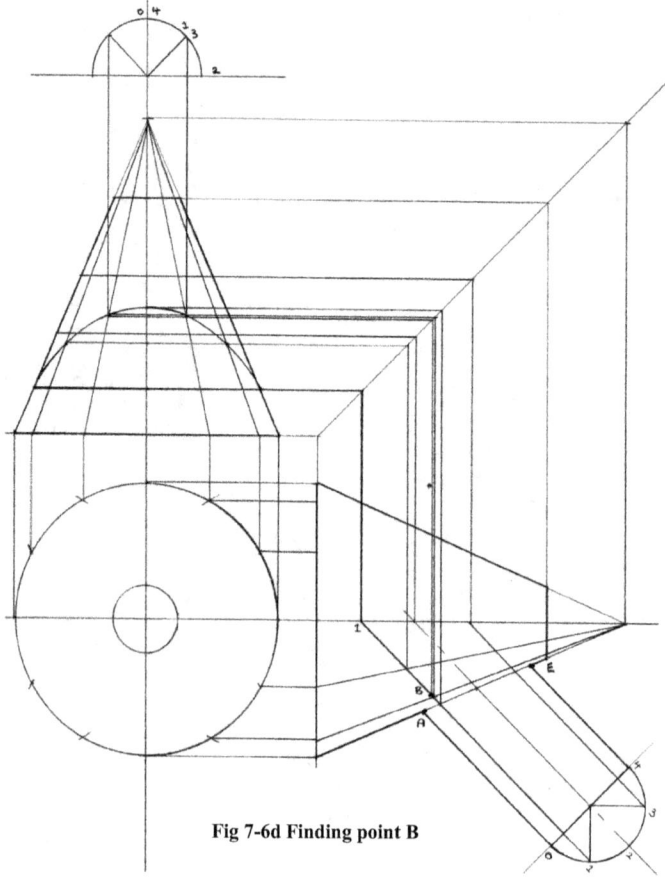

Fig 7-6d Finding point B

Line **1** produces 4 intersection points with the lines from the base of the cone to the apex, one is

at the centre line, one is at the full diameter and the other two are on the radial lines.

Each of these points is transferred as horizontal lines to intersect with the radial lines drawn in the front elevation to produce points marking a sectional cut. These points are connected with a fluid curved line. A half-diameter of the pipe is drawn on the centre line above the front elevation and divisions identical to those in the side elevation are marked. The corresponding points are projected vertically down to intersect with the cone section.

This intersection point is then transferred to the side elevation to intersect line **1 at point B.**

If greater accuracy is required the inclined plane **1 to the intersection point with the outside of the cone** (the line between point 1 on the centre line and the outside edge of the cone) can be imagined as a sectional cut and projected at 90° to be drawn as a true length with the **3** chords **AB, CD and EF** together with point **G** from the front elevation drawn in their correct position as in **fig 7-7d**.

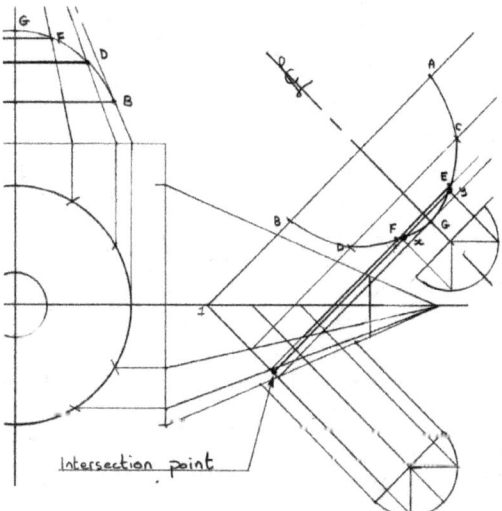

The corresponding points are projected onto the true shape of the section and then transferred to the side elevation to give the required intersection point **B**.

Fig 7-7d Inclined plane

In **fig 7-8d** line **2** produces 4 intersection points with the radial lines.

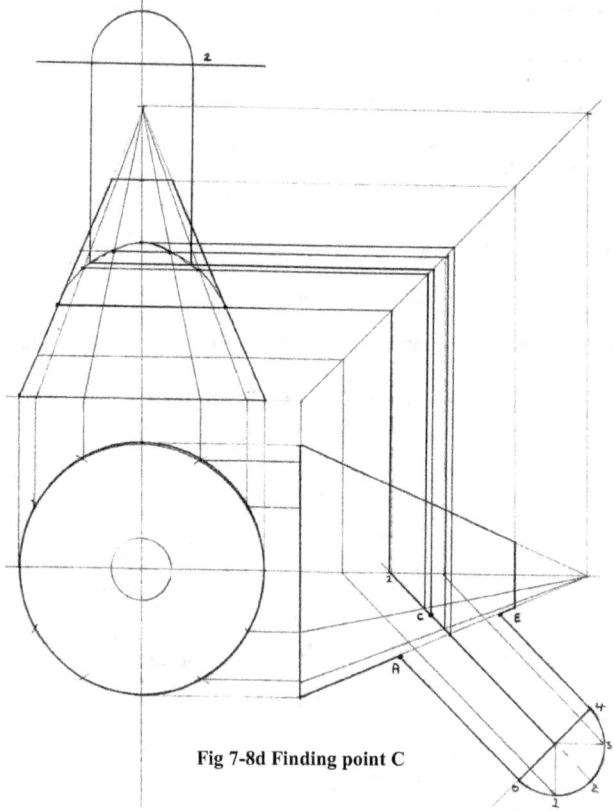

Fig 7-8d Finding point C

As before these points are transferred as horizontal lines to intersect with the radial lines in the front elevation. The intersection points are drawn as a sectional cut and the corresponding lines are projected from above to intersect with the cone section.

This intersection point is transferred to the side elevation to intersect line **2 at point C.**

In **fig 7-9d** line **3** produces 4 intersection points with the radial lines.

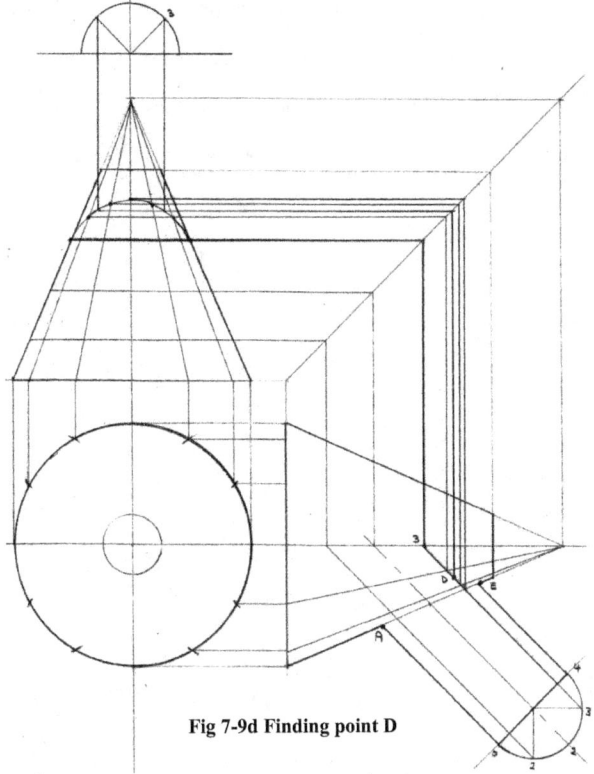

Fig 7-9d Finding point D

Again these points are transferred as horizontal lines to intersect with the radial lines in the front elevation. The intersection points are drawn as a sectional cut and the corresponding lines are projected from above to intersect with the cone section.

This intersection point is transferred to the side elevation to intersect line **3 at point D.**

With all intersection points marked on the side elevation, they can now be transferred to the other two views to assist with the production of the templates. From the side elevation all points are transferred as horizontal lines to the front elevation where they are intersected from above by the pipe division lines. The points in the plan view are found at the intersections of the points horizontally from the side view and vertically from the front elevation as in **fig 7-10d.**

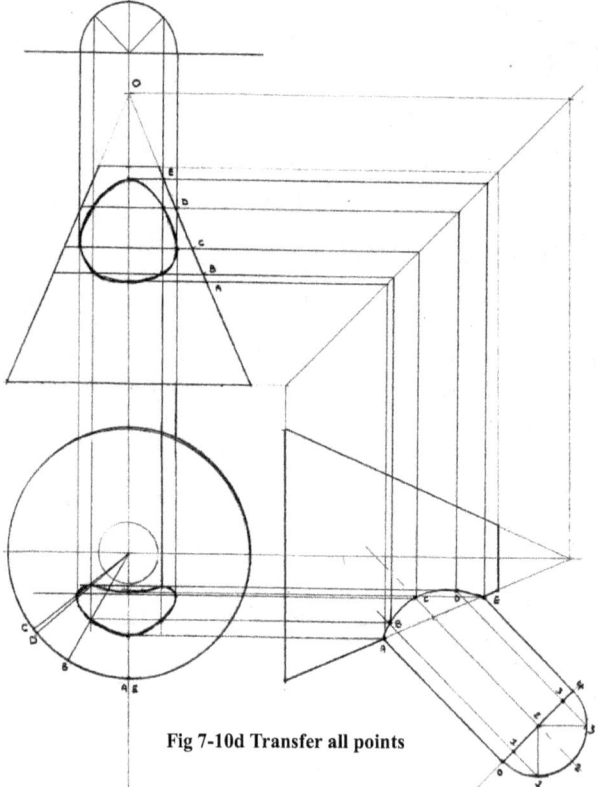

Fig 7-10d Transfer all points

These points are marked on the right hand side of the front elevation as **A, B, C and D.**

Radial lines are drawn through the intersection points in the plan view to produce the points **A, B, C and D** on the circumference.

The cone and pipe templates can now be produced.

Cone template.

At the top of **fig 7-11d** the cone is marked out as in chapter 5.

A centre line is drawn and at the point where it intersects the circumference, the starting point for **A and E** is found.

All sizes are taken directly from **fig 7-10d.**

From this point with the dividers or trammel points set at the chord distance **A -B** in the plan view, point **B** is marked on each side.

With the dividers or trammel points set at the chord distance **B - D** in the plan view, point **D** is marked from **B** on each side.

With the dividers or trammel points set at the chord distance **C -D** in the plan view, point **C** is marked from **D** on each side.

A line is drawn from each point back to the apex.

From the apex marked **O** on the front elevation the distance to each point on the right hand sloping side of the cone is marked as an arc on the template from the apex, and the relevant intersection points are plotted.

A fluid curved line drawn between all points completes the template.

Pipe template.

The pipe template at the bottom of **fig 7-11d** is marked using sizes taken directly from **fig 7-10d.**

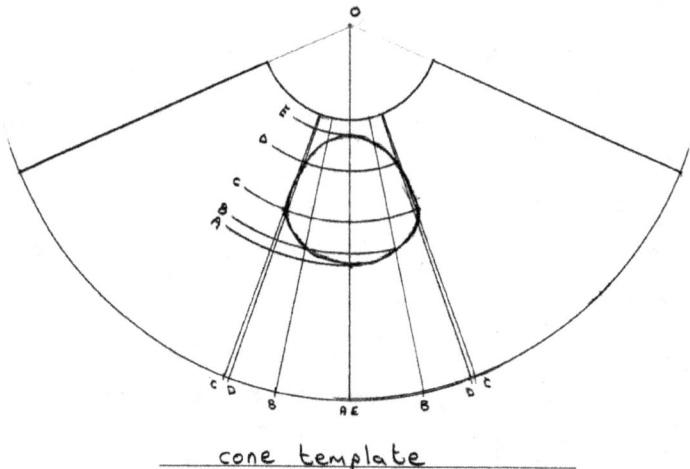

cone template

Fig 7-11d Intersection templates

pipe template

The diameter of the pipe multiplied by π will give its circumference, which will be the length of

the template.

The circumference divided by 8 will give the distance between the perpendiculars on the template.

From the end of the pipe on the side elevation, starting at **O** the distance **O - A** is marked on each end of the template.

All distances are now transferred to the template, i.e., **1 - B, 2 - C, 3 - D and 4 - E.**

A fluid curved line is drawn between all points to complete the template.

In this example sectional cuts through the cone needed to be plotted before intersection points could be established. Every intersection will need to be determined on its own merits, but, with the lessons learned in this chapter, the boilermaker will gain the skills necessary to make the informed decisions he will be called upon to make. Obviously not every intersection has been shown, but the principles shown in this chapter can be applied in many forms and are intended to give the boilermaker a good basic knowledge of those principles.

CHAPTER 8

Ellipse

8a Explanation of.

An ellipse is the figure produced when a circle is projected onto an inclined surface or a curved surface. From a view point perpendicular to the circle, the circle appears round. From a view point perpendicular to the inclined surface the circle takes on the appearance of an ellipse, it has the same diameter as the circle in one plane, but it is elongated (as a sectional cut on a pipe) or shortened (as a tilted flat disc) in the other.

Example.

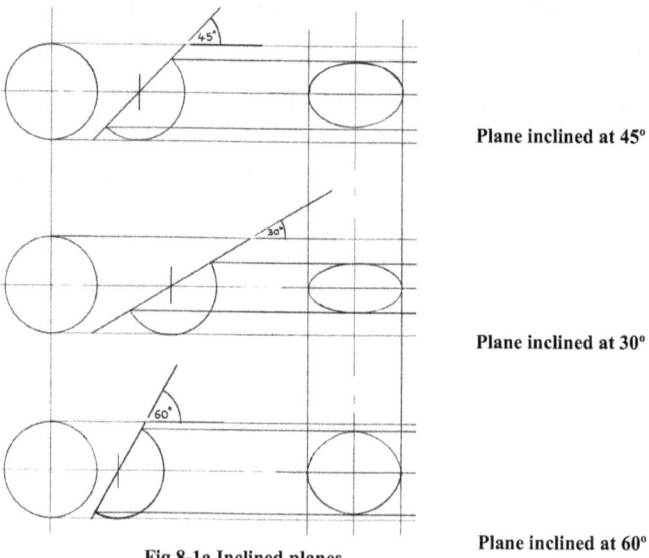

Plane inclined at 45°

Plane inclined at 30°

Plane inclined at 60°

Fig 8-1a Inclined planes

Fig 8-1a shows a circle that has produced an ellipse with three different shapes by altering the angle of inclination of the surface it is projected onto. The diameter in one plane stays the same but varies in the other plane. It can be seen that the greatest angle of inclination (furthest away from 90°) produces the longest ellipse and the smallest (closest to 90°) inclination produces an ellipse closer to a circle.

There are a few different ways to mark an ellipse; the only ones this chapter will be concerned with are those where the value of both the greater and lesser axis are known.

An ellipse has two diameters; **the lesser and greater axis.**

The lesser axis is the diameter of the circle being used, which will usually be the outside diameter of the pipe or cone intersecting with the inclined or curved surface.

The greater axis is the amount of elongation involved in the particular work-piece.

8b Mark out on a flat surface.

There are two very efficient methods of drawing an ellipse when both diameters are known.

The first one is to construct a rectangle around the two diameters.

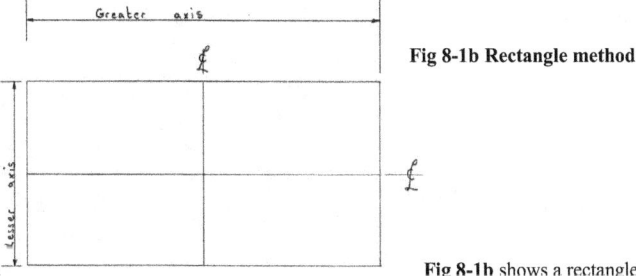

Fig 8-1b Rectangle method

Example. Fig 8-1b shows a rectangle with

the longest side being the same measurement as the greater axis of an ellipse and the shorter side equal to the lesser axis.

A centre line is drawn through each side dividing the rectangle into four equal parts.

In **fig 8-2b** the longest centre line has been divided into **8 equal parts** and the two short sides of

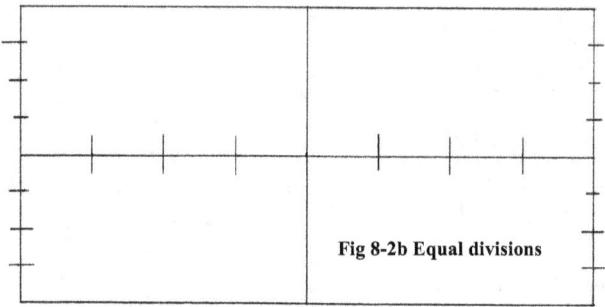

Fig 8-2b Equal divisions

the rectangle also divided into **8 equal parts.** Any number can be used, as long as it is the same

for each axis.

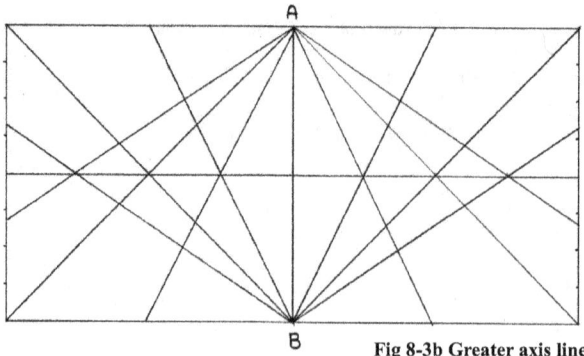

In **fig 8-3b**   Fig 8-3b Greater axis lines

lines have been drawn from points **A and B** through the division points on the longest centre line

to the outside of the rectangle.

In **fig 8-4b** lines have been drawn from points **A and B** to the division points on the short sides of the rectangle, thereby producing intersection points which are joined with a fluid curved line to complete the template.

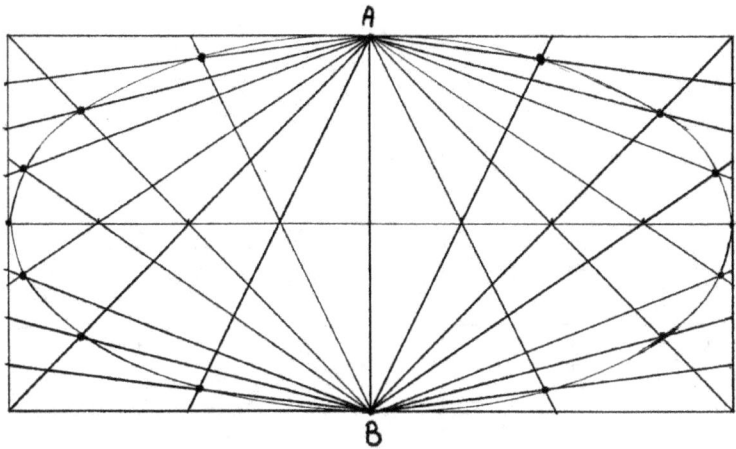

Fig 8-4b Intersection points

The second and simpler method is to use the two diameters projected onto each other which can be done two ways.

Example 1.

Fig 8-5b shows the two diameters as half circles divided into a number of equal parts. Any number can be used as long as it is the same number for each diameter.

The projected intersection points plot the shape of the ellipse.

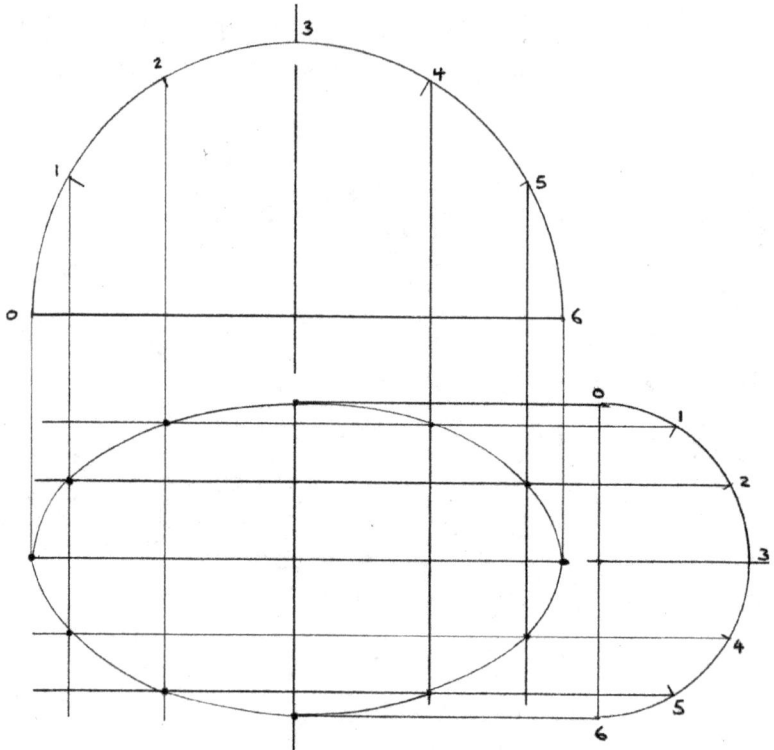

Fig 8-5b Projected intersection points

Example 2.

Fig 8-6b shows the two diameters as concentric circles divided into a number of equal parts, shown as radial lines.

Vertical lines are drawn from the division points on the large circle (greater axis).

Horizontal lines are drawn from the division points on the small circle (lesser axis).

The intersection points plot the shape of the ellipse.

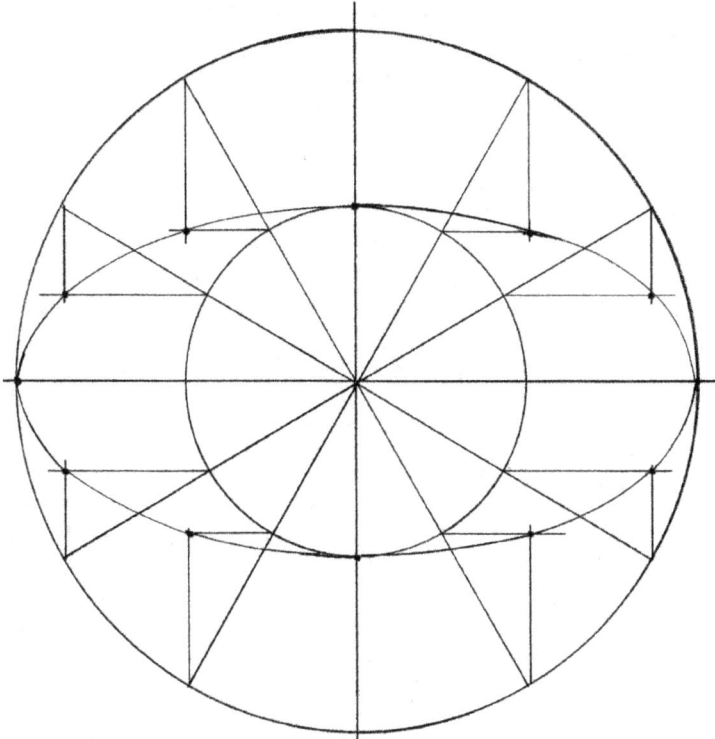

Fig 8-6b Concentric circles

8c Template for curved surface.

The ellipse template for a curved surface is made in exactly the same way as that for a flat surface. The difference is that the greater axis will need to be determined first. This is done with the knowledge of the orientation of the pipe on the side of the vessel (position in degrees) and its

attitude to the vessel (angle at which it meets the vessel).

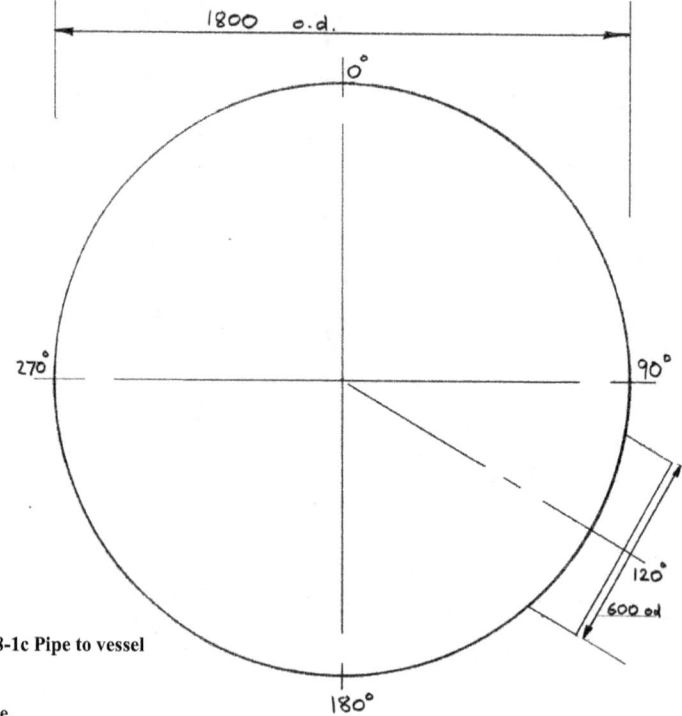

Fig 8-1c Pipe to vessel

Example.

A pipe **600** outside diameter is to be welded into a vessel which has an outside diameter of **1800**

at the **120°** position and is to be central with the vessel (the pipe must point to the vessel centre).

Fig 8-1c shows the pipe in position. The lesser axis is **600** which is the diameter of the pipe.

The greater axis is determined by calculating the angles where the outside of the pipe meets the

vessel.

Fig 8-2c shows a quarter section of the vessel with the pipe in position and lines drawn back to the centre from the points where the outside of the pipe meets the vessel. It can be seen that a chord drawn between these two points makes two identical right angled triangles.

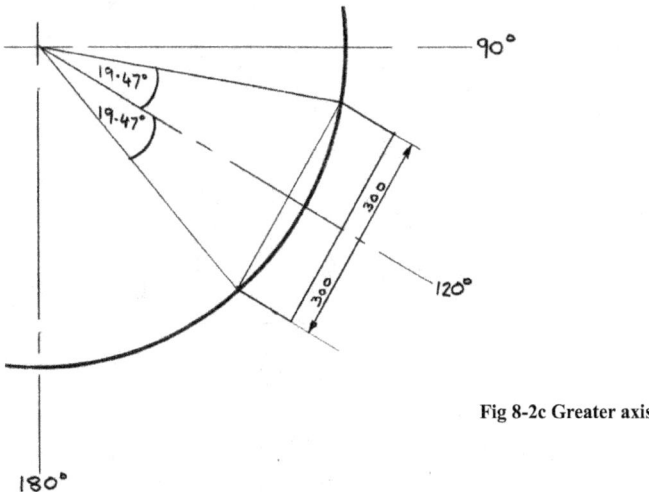

Fig 8-2c Greater axis

The angle at the centre is determined by $\sin = \dfrac{opp}{hyp}$ ...... $\sin = \dfrac{300}{900}$ .... angle = **19.47°.**

The angle of arc the nozzle is bounded by is..... 19.47 x 2 = **38.94°.**

1° of arc is determined by $\dfrac{pi\, x\, diameter}{360}$ .... $\dfrac{pi\, x\, 1800}{360}$ .... **1° = 15.7.**

The length of the angle of arc is determined by .... **38.94 x 15.7 = 611.3.**

The greater axis is **611.3.**

The template for the hole in the vessel is made as if the ellipse was on a flat surface and put into place on the curved surface of the vessel for marking around.

© The Mathematics of Boilermaking by Jim Draper 2017
137

8d Template with double elliptical value.

Fig 8-1d shows the same nozzle on the same vessel in the same position but with a different attitude. The pipe is now horizontal with the vessel.

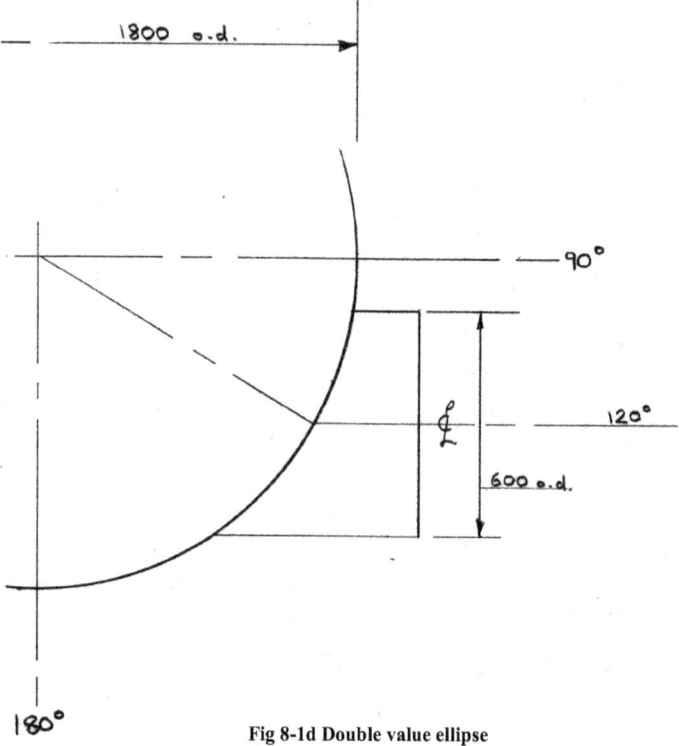

Fig 8-1d Double value ellipse

It can be seen that the bottom half of the pipe has a longer length of arc than the top half, which means the two halves will have a different greater axis. The lesser axis always stays the same.

Fig 8-2d shows the lines drawn back to the centre as before.

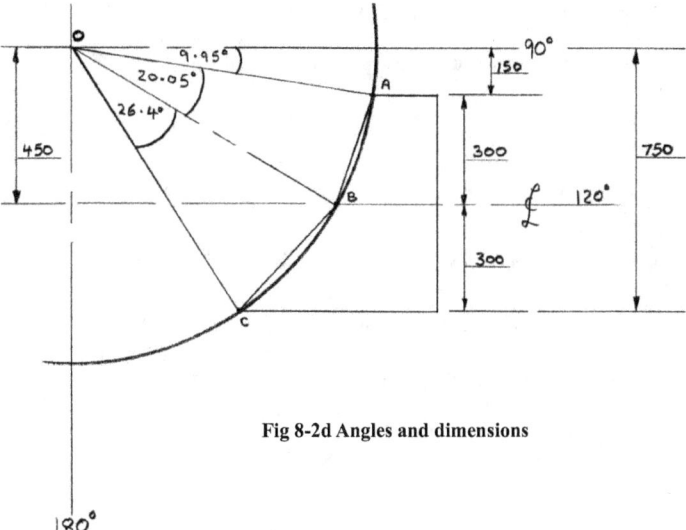

Fig 8-2d Angles and dimensions

Two chords will be needed to form two triangles for this exercise, one from the centre **B** to the top of the pipe at **A** and the other from the centre **B** to the bottom of the pipe at **C**.

The distance below the **90° line** for points **A, B and C** will need to be known to find the apex angles of two triangles **AOB and BOC**.

The distance for point **B** can be found by .... 120° - 90° = **30°**.

opp – sin x hyp........ opp – sin30 x 900 – **450**.

139

The distance for point **A** can be found by .... 450 - 300 = **150.**

The distance for point **C** can be found by .... 450 + 300 = **750.**

The angle at the apex between the **90° line** and point **A** can be found by.....

$$\sin = \frac{opp}{hyp} \ \cdots \ \sin = \frac{150}{900} = \mathbf{9.95°.}$$

The angle between the **90° line** and point **B** = **30°.**

The angle between the **90° line** and point **C** can be determined by $\sin = \dfrac{750}{900} = \mathbf{56.4°.}$

The apex angle for the triangle **AOB** can be found by........ 30° - 9.95° = **20.05°.**

If this angle is multiplied by the length of 1° of arc

it will give the length of arc between A **and B.**

15.7 x 20.05 = **314.78.**

The apex angle for triangle **BOC** can be found by .... 56.4° - 30° = **26.4°.**

The length of arc between **B and C** is found by .... 15.7 x 26.4 = **414.48.**

The two greater axis **radii** needed for this ellipse are **314.78 and 414.48.**

The two greater radius values can now be used to make the template needed to place on the vessel and mark the cut out needed to let the pipe enter at the correct attitude.

Fig 8-3d shows the template needed to mark this ellipse with the two greater axis **radii** being used.

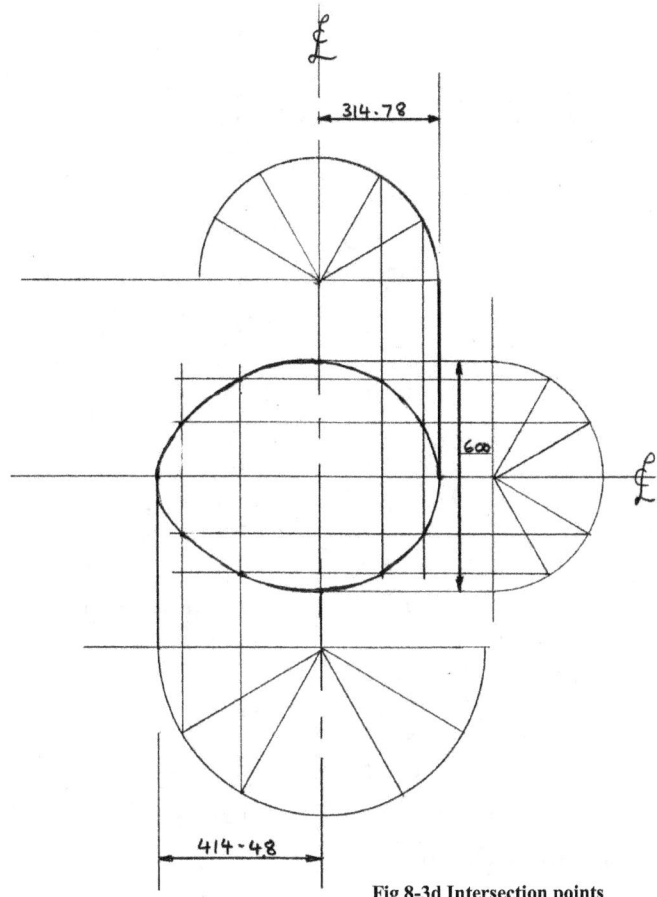

Fig 8-3d Intersection points

CHAPTER 9

Spirals

9a Helix.

A helix is formed when something is wound around a cylinder while it is moving along that cylinder.

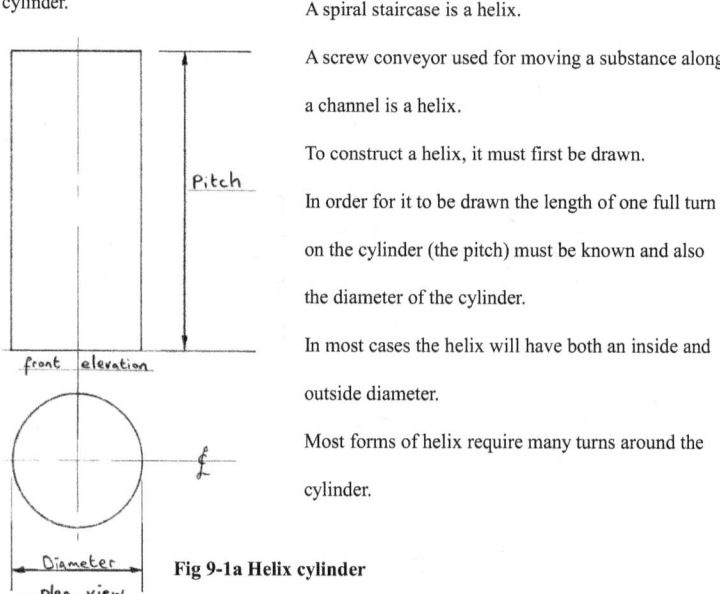

A spiral staircase is a helix.

A screw conveyor used for moving a substance along a channel is a helix.

To construct a helix, it must first be drawn.

In order for it to be drawn the length of one full turn on the cylinder (the pitch) must be known and also the diameter of the cylinder.

In most cases the helix will have both an inside and outside diameter.

Most forms of helix require many turns around the cylinder.

Fig 9-1a Helix cylinder

One full pitch will be shown in the example; the process is simply repeated until the required amount of turns or pitches has been reached.

Example.

Fig 9-1a shows a plan view of a cylinder with the front view of the cylinder above it. The height of the cylinder will be one pitch or one revolution.

When marking out a right hand helix, start at the left side and number towards the right. With a left hand helix the opposite applies.

In **fig 9-2a** the cylinder has been divided into **12 equal parts** and the divisions have been projected up onto the front elevation.

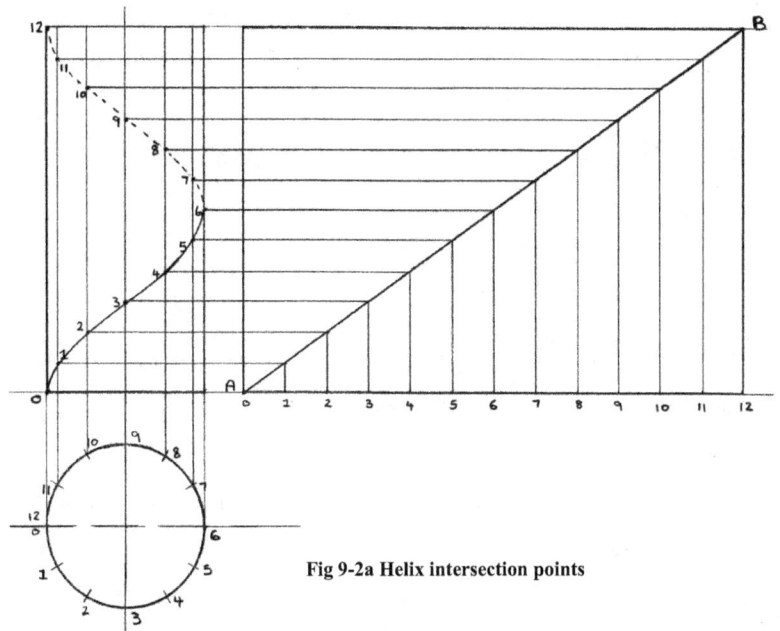

Fig 9-2a Helix intersection points

The drawing to the right represents the cylinder opened out to a circumference length and also divided into **12 equal parts.**

A line has been drawn between the opposite corners **A and B.**

The **12** division lines are drawn vertically to meet line the **AB,** where they are drawn horizontally across to the front elevation to intersect with lines drawn from the plan view and produce the intersection points that plot the helix.

The points are joined with a fluid curved line to show the helix.

Points **1-6** are joined with a solid line as they are seen at the front of the front elevation.

Points **6-12** are joined with a broken line as they are hidden at the back of the front elevation.

From this example it can be seen that the length of the cylinder that the helix travels around is

πD, or, one circumference.

However the length of the helix is equal to the length of the line **AB**.

Which can be found by using $a^2 + b^2 = c^2$.

The true length of the helix is ....$\sqrt{\text{circumference}^2 + \text{pitch}^2}$.

A helix can have more than one diameter, an example would be a spiral stair case, where the end of the stair furthest from the cylinder will have a greater diameter than the end fixed to the cylinder.

Both would share the same pitch on the cylinder but the outer spiral would be longer because of its greater diameter.

Example.

Fig 9-3a shows a double helix similar to one that might be used on a spiral stair case.

The outer spiral is plotted using the outer circumference on the right and the outer diameter division spaces on the front elevation.

The inner spiral is plotted using the inner circumference on the left and the inner diameter division spaces on the front elevation.

Thus; producing two spirals on the same pitch.

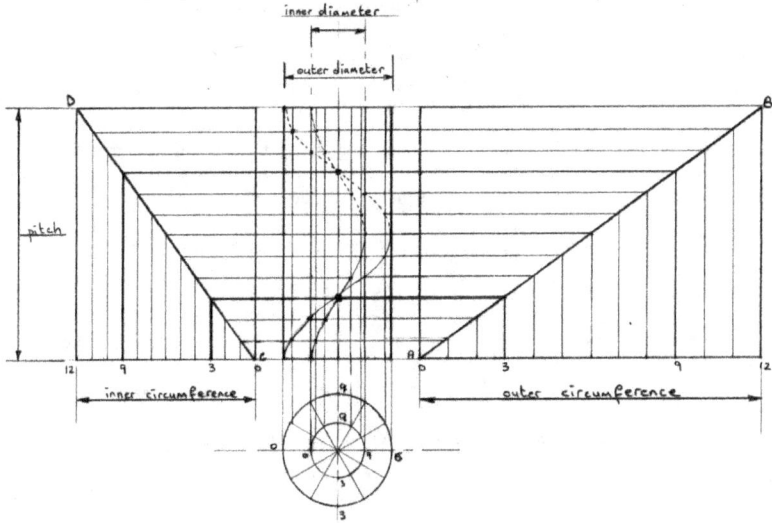

Fig 9-3a Double helix

It can be seen that the two spirals cross each other at the same point on centre lines **3 and 9** on the front elevation.

It can further be seen that the horizontal lines from both the inner and outer circumference line up with each other on the front elevation. This is because from either side the **pitch** is still divided into **12 equal** parts.

The outer helix will be the length of the line **AB.**

The inner spiral will be the length of the line **CD.**

Four spirals can be plotted on one pitch.

An example would be a rectangular spiral chute delivering material to a point directly below its starting point, where it is not practical to let the material fall straight down. The chute will also act as a guide to give direction to the material.

The spiral will slow the rate of fall and also direct the material when it reaches the desired height. The two outer spirals will be identical although separated by the depth of the chute and the two inner spirals will be identical although separated by the depth of the chute.

Therefore the pitch will need to be drawn twice separated by the depth of the chute.

Example.

In this example, for simplicity, the depth of the chute will be equal to one twelfth of the height of the pitch, **or one division.** This simply means that an extra division will be added to the top of the pitch length, and the second set of spirals will start and end one division up from the first set of spirals.

Fig 9-4a shows a rectangular chute with the same inner and outer diameters as the previous example and its depth is equal to one twelfth of the pitch.

This exercise is drawn out exactly the same as in the last exercise with the exception being that there is a second set of spirals beginning and ending one division above the first set.

The depth of the chute can be fixed on a pair of dividers or set of trammels and simply marked on the vertical lines on the front elevation above the points used for the first two spirals drawn in **fig 9-3a.**

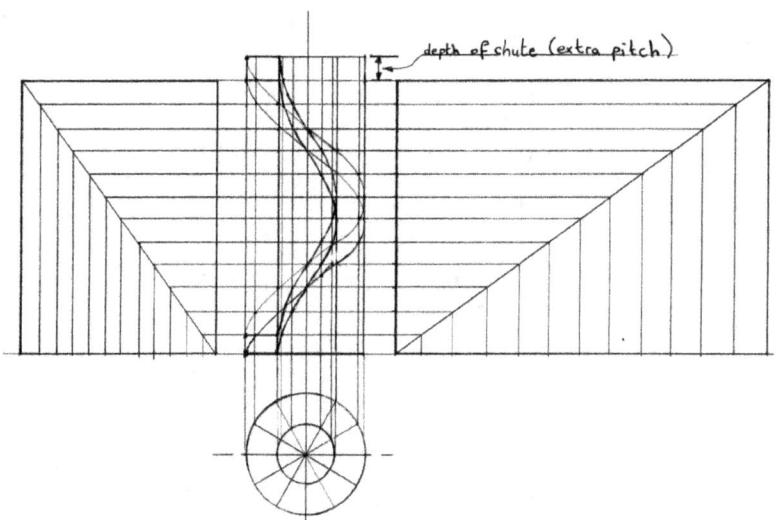

depth of chute (extra pitch)

Fig 9-4a Helical rectangular chute

Extreme care must now be taken so as not to become confused amongst the many lines that will

be seen.

Fig 9-5a shows the helical rectangular chute enlarged and with shading to give a 3-D effect making the finished product easier to see.

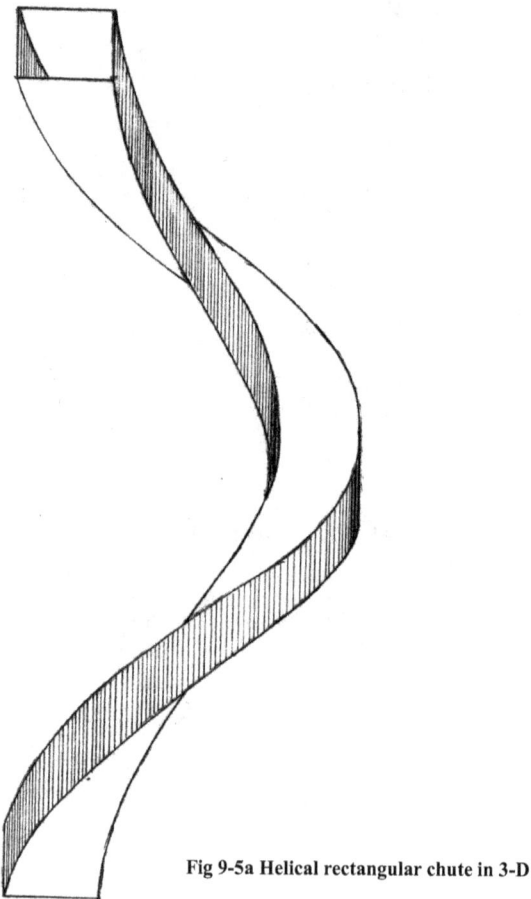

Fig 9-5a Helical rectangular chute in 3-D

9b Regular screw conveyor flight.

Screw conveyors are constructed one pitch (called a flight) at a time and then welded together to produce the desired length or number of pitches required.

The flights make up the helix which moves or conveys the material inside a chute or trough. Each flight is one pitch.

Fig 9-1b shows a regular screw conveyor with **3** pitches or flights. Each flight is welded to the one next to it at the point marked **X.**

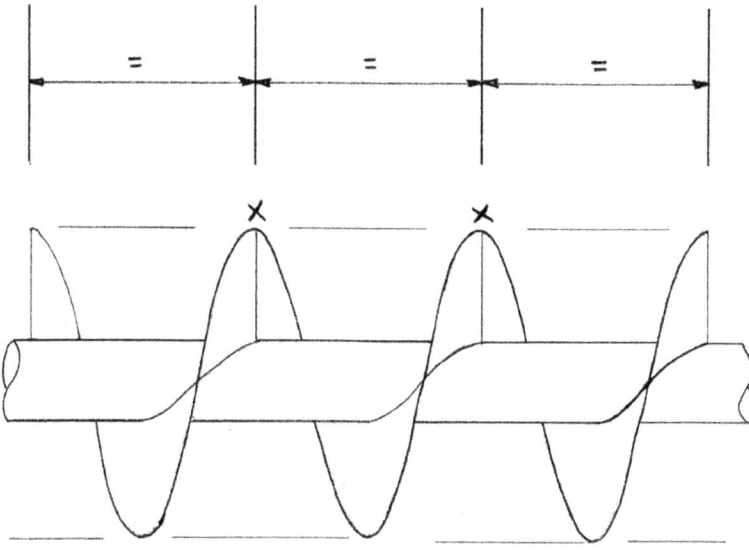

Fig 9-1b Regular screw conveyor

It is a regular screw conveyor because all flights have the same inner and outer diameter and the same pitch.

To make a screw conveyor flight the lessons learned at the beginning of the chapter will be applied.

It was seen that the length of the helix was the **square root** of (circumference² + pitch²).

With a screw conveyor flight there is an inner and an outer circumference sharing the same pitch as in **fig 9-2b** showing one **formed flight**, therefore two calculations need to be made.

The true inner and outer circumferences will need to be determined in order to make the template for the pre-formed flight.

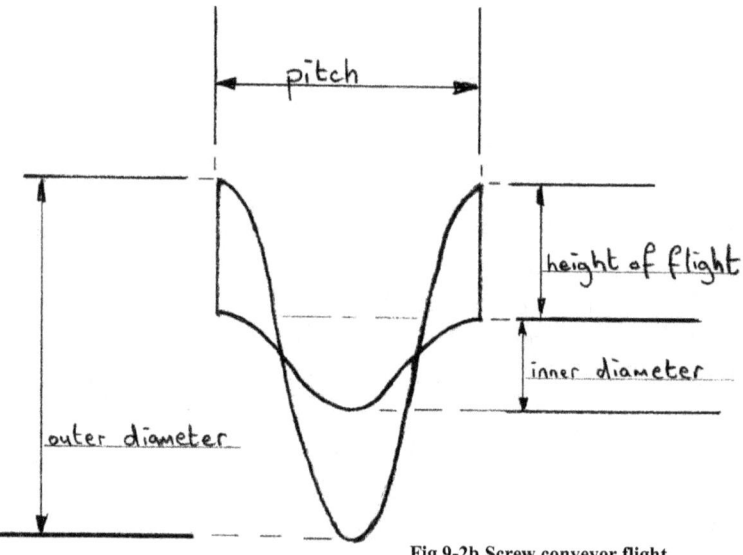

Fig 9-2b Screw conveyor flight

These two sizes will then be used together with the height of the flight to calculate the angle of the cut-out that will be necessary as shown in **fig 9-3b,** depicting a pre-formed flight meaning it is still a flat disc.

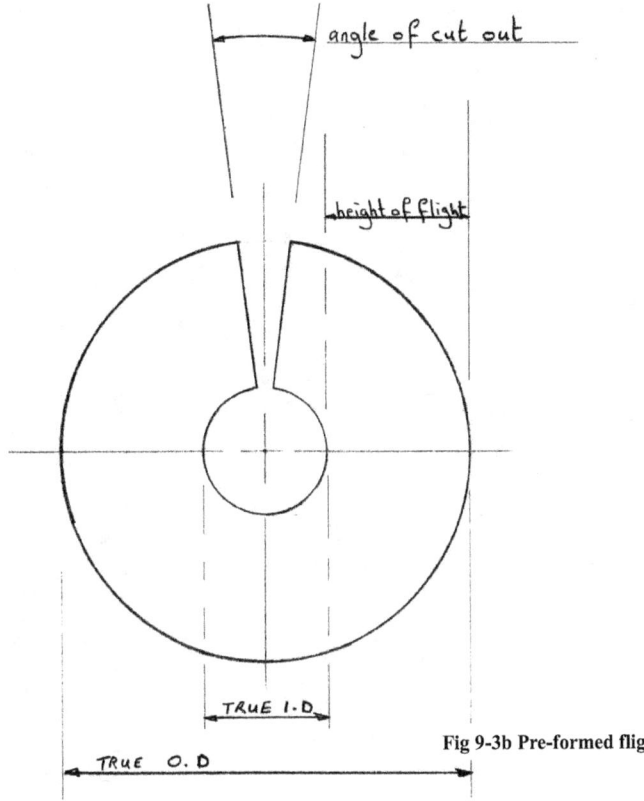

angle of cut out

height of flight

TRUE I.D

Fig 9-3b Pre-formed flight

TRUE O.D

It can be seen from **fig 9-3b** that the **od and id** of the template are shown as **true** sizes, these will

need to be determined along with the angle of the cut-out as follows :

Step 1. Determine the height of the flight.

Step 2. Determine outer helix circumference.

Step 3. Determine inner helix circumference.

Step 4. Determine the true inside radius.

Step 5. Determine the true outside radius.

Step 6. Determine the angle of the cut-out.

Example.

Fig 9-4b shows a screw conveyor with:

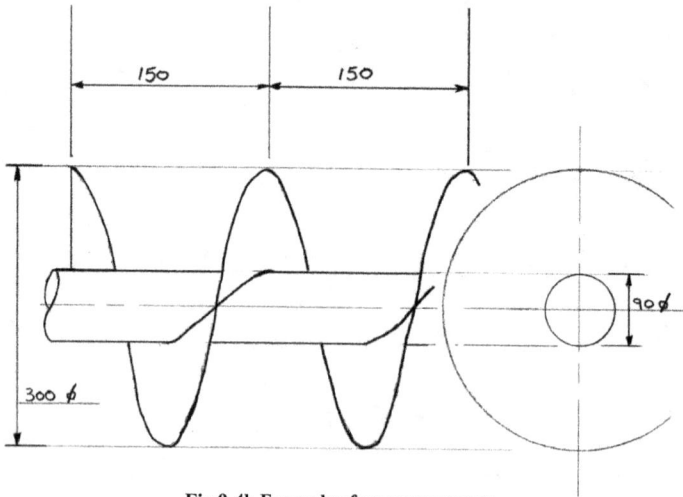

Fig 9-4b Example of screw conveyor

od = 300 .... id = 90 .... pitch = 150.

Step 1. Height of flight = $\dfrac{od - id}{2}$ = **105.**

Step 2. Outer helix circumference = $a^2 + b^2 = c^2$ ... (od x π)² + pitch² = helix circumference².

$\sqrt{(300 \times \pi)^2 + 150^2}$ = **954.33.**

Step 3. Inner helix circumference = **(id x π)² + pitch² = helix circumference².**

$\sqrt{(90 \times \pi)^2 + 150^2} = 320.$

Step 4. The true inside radius is found by constructing a right angled triangle with the flight height as the perpendicular side and then calculating the apex angle which is used to determine the radius by constructing two triangles one above the other.

In **fig 9-5b** two triangles are shown one above the other.........

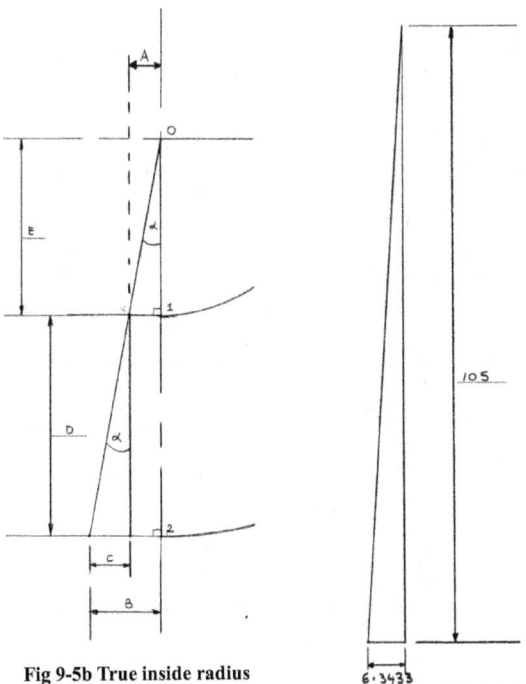

Fig 9-5b True inside radius

Fig 9-6b Determine the angle

0 is the centre of the flight template.

0-1 is the true inside radius.

0-2 is the true outside radius.

A is the inner helix circumference divided by 100.

B is the outer helix circumference divided by 100.

C is B minus A.

D is the height of the flight.

This information is used to construct a right-angled triangle to determine the angle needed to find the true inside radius.

The base of the triangle is the relationship between the outer and inner helix circumferences when both are divided by 100. Thus;

$$\frac{outer\ helix\ circumference}{100} = \frac{954.33}{100} = \textbf{9.5433.}$$

$$\frac{inner\ helix\ circumference}{100} = \frac{320}{100} = \textbf{3.2.}$$

The inner is now subtracted from the outer .... 9.5433 - 3.2 = **6.3433.**

This is the base of the triangle as shown in fig 9-6b.

The height of the flight is the vertical side of the triangle.

With two sides of the triangle known the angle opposite the base can be calculated............

$$\tan = \frac{opp}{adj} \quad \tan = \frac{6.3433}{105} = \textbf{3.457°.}$$

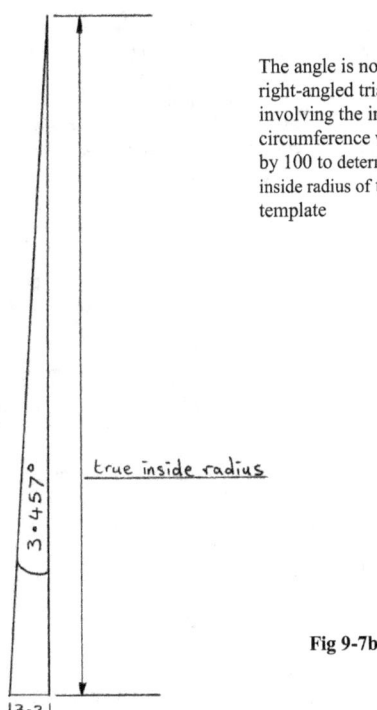

The angle is now used in a right-angled triangle, involving the inner helix circumference when divided by 100 to determine the true inside radius of the flight template

Fig 9-7b Using the angle

This angle is now used in a right angled triangle together with the inner helix circumference when divided by 100 as shown in **fig 9-7b.**

The true inside radius is the perpendicular of the triangle.....

$$adj = \frac{opp}{\tan} \qquad adj = \frac{3.2}{\tan 3.457} = 52.97.$$

Step 5. The true outside radius is found by adding the height of the flight to the true inside radius......

105 + 52.97 = **157.97.**

Step 6. The angle of the cut-out is the relationship between the true circumference of the flight and the circumference of the true diameter which are two different sizes.

The true circumference of the flight is the outer helix circumference which is **954.33.**

The circumference of the true diameter is.... (true outside radius x 2) x π...........

157.97 x 2 = 315.94.......... 315.97 x π = **992.55.**

The relationship between the two can be found by dividing the lesser number by the greater number.....

$$\frac{954.33}{992.55} = \textbf{0.961.}$$

The same calculation is done with the inner circumference as a way of checking accuracy as .... **both calculations must produce the same answer.**

The true circumference of the inner helix is **320.**

The circumference of the true inner diameter is (true inside radius x2) x π................

52.97 x 2 = 105.94.......... 105.94 x π = **332.82.**

$$\frac{320}{332.82} = \textbf{0.961.}$$

This number represents the percentage of a full circle needed for the flight.

To convert it to an angle it is multiplied by the total number of degrees in a circle..........

0.961 x 360 = **345.96°.**

The angle of the cut-out is found by subtracting this figure from 360.

360 - 345.96 = 14.04°.

All of the information is now transferred to a flight template as in **fig 9-8b** and a number of identical flights are produced and welded to each other according to requirement.

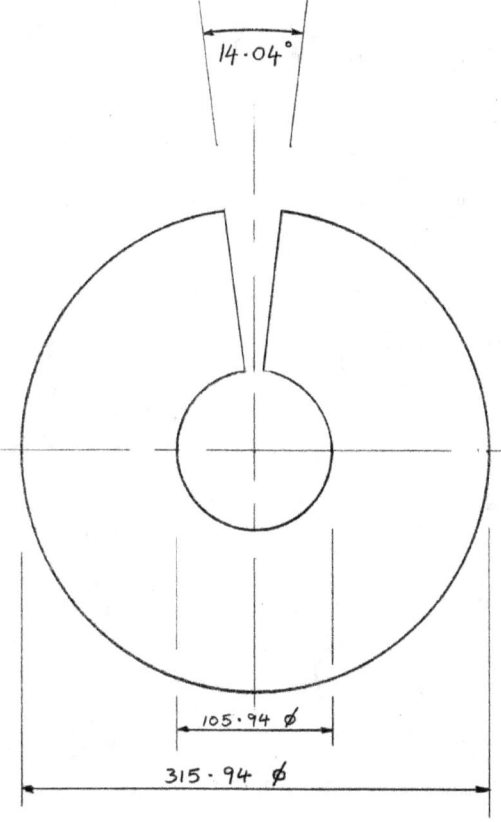

Fig 9-8b Flight template

True outside diameter = (157.97 x 2)........ **315 94.**

True inside diameter = (52.97 x 2) .... **105.94.**

Cut-out angle =.. **14.04°.**

9c Irregular screw conveyor flights.

An irregular screw conveyor is one where; the pitch, the outside diameter or the inside diameter does not remain constant.

A screw conveyor with a pitch that gets bigger or smaller along its length is an irregular screw conveyor.

Screw conveyor flights on a cone are irregular.

Screw conveyor flights where the outside diameter increases or decreases along its length are irregular.

Example 1.

Fig 9-1c shows a screw conveyor with a constant inside and outside diameter with a varying pitch.

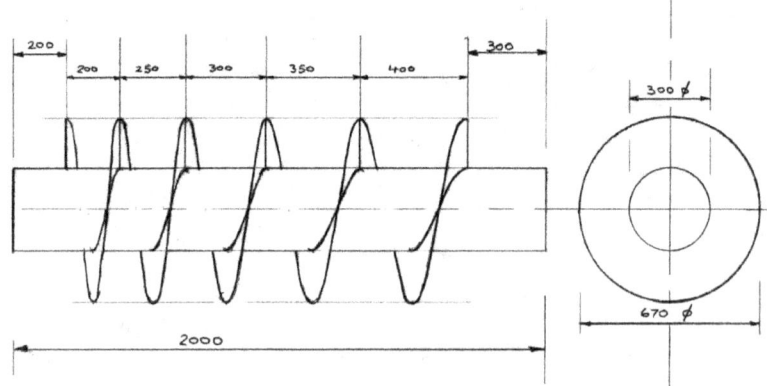

Fig 9-1c Irregular screw conveyor

This example can be dealt with as a regular screw conveyor with each flight template being made separately as in section **9b.** The three pieces of information needed for each flight are...

outside diameter (which is the same for all flights).

inside diameter (which is the same for all flights).

pitch (which is different for each flight).

Example 2.

Fig 9-2c shows a screw conveyor with a constant outside diameter and pitch, built on a cone.

As in all irregular screw conveyors each flight is unique and a template is made for each and

every one.

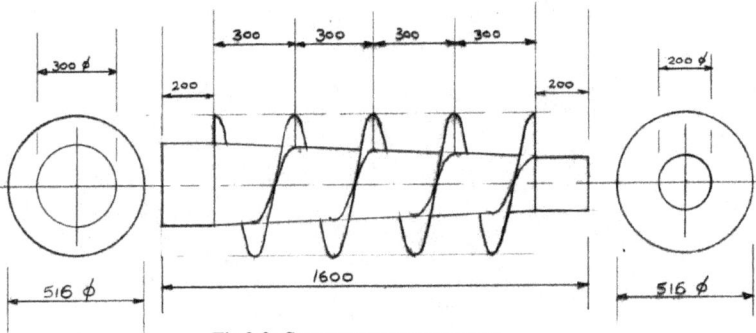

Fig 9-2c Screw conveyor on a cone

The same three pieces of information are still required.................

outside diameter (which is the same for all flights).

inside diameter (which is different for each flight).

pitch (which is the same for all flights).

The first thing is to determine the inside diameter at the points where each flight begins and ends.

This can be found using the knowledge of the length of the cone together with the diameter at

each end.

In this example the length of the cone is the pitch multiplied by 4 .... 300 x 4 = **1200.**

The widest end of the cone is **300 diameter.**

The narrowest end of the cone is **200 diameter.**

The difference in the two diameters is 300 - 200 = **100.**

The pitch will increase or decrease its inside diameter by 4 equal parts therefore 100 divided by 4

= **25.**

Starting with a diameter of 300 each division narrows by 25.

in **fig 9-3c** the divisions have been numbered 0-4.

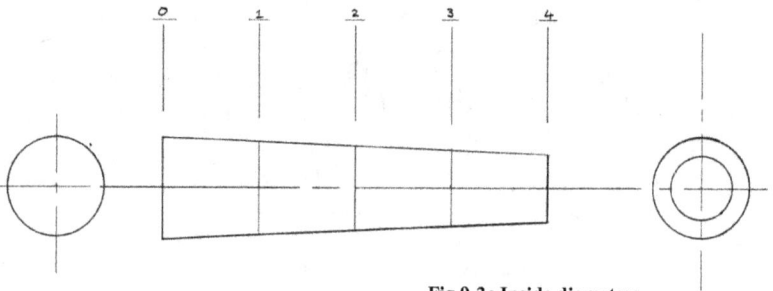

Fig 9-3c Inside diameters

The diameter at point 0 is **300.**

The diameter at point 1 is **275.**

The diameter at point 2 is **250.**

The diameter at point 3 is **225.**

The diameter at point 4 is **200.**

The four flights each require their own template.

The flights will be marked **A, B, C and D.**

Flight **A** has an inside diameter that starts as **300** and finishes as **275.**

Flight **B** has an inside diameter that starts as **275** and finishes as **250.**

Flight **C** has an inside diameter that starts as **250** and finishes as **225.**

Flight **D** has an inside diameter that starts as **225** and finishes as **200.**

Each template is made as in section **9b.**

The same three pieces of information will be required, namely; inside diameter, outside diameter and pitch. For each template the largest inside diameter will be used as a starting point.

Example.

With reference to **fig 9-2c.**

Template **A.**

The height of the flight is $\dfrac{516 - 300}{2}$ = **108.**

Outer helix circumference = $\sqrt{(516 \times \pi)^2 + 300^2}$ = **1648.58.**

Inner helix circumference = $\sqrt{(300 \times \pi)^2 + 300^2}$ = **989.07.**

Pitch = **300.**

Find the required angle....................

$\dfrac{1648.58}{100} - \dfrac{989.07}{100}$ = **6.5951..........**

$\tan \dfrac{opp}{hyp} = \dfrac{6.5951}{108}$ .... angle = **3.494°.**

Find the true inside radius...........................

$adj = \dfrac{opp}{tan} = \dfrac{9.89}{tan\,3.494}$ = **161.97..........**

© The Mathematics of Boilermaking by Jim Draper 2017

161

True id = 161.97 x 2 = **323.94** .... circumference = 323.94 x π = **1017.68.**

True outside radius = 161.97 + 108 = **269.97.............**

True od = 269.97 x 2 = **539.94....................................** circumference = 539.94 x π = **1696.27.**

Find the cut-out angle...............................

$\dfrac{1648.58}{1696.27}$ = 0.9718 $\dfrac{989.07}{1017.68}$ = 0.9718   .... 0.9718 x 360 = **349.848°...**

360 - 349.848 = **10.152°.**

The template for flight A is shown in **fig 9-4c.**

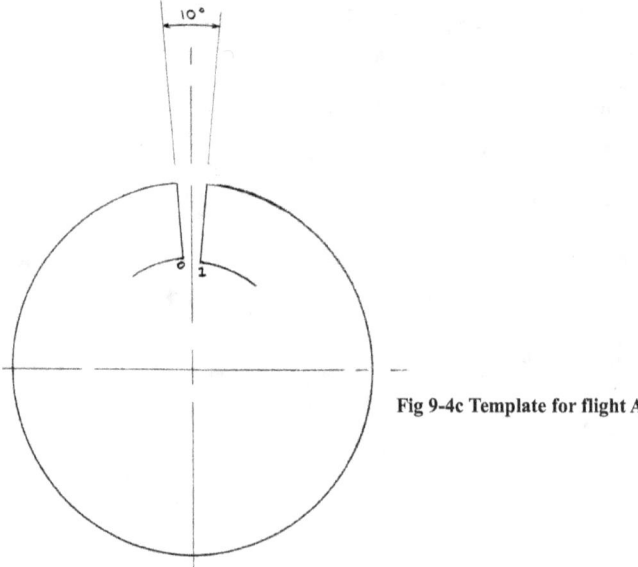

Fig 9-4c Template for flight A

The cut-out angle of **10°** has been marked and the flight heights at points **0** and **1** have been marked from the outside circumference.

The flight height at position **0** is **108.**

Since the diameter decreases by **25** at each point, the flight height will increase by half this amount which is **12.5.**

The flight height at position **1** is .... 108 + 12.5 = **120.5.**

The flight height at position **2** is..... 120.5 + 12.5 = **133.**

The flight height at position **3** is..... 133 + 12.5 = **145.5.**

The flight height at position **4** is..... 145.5 + 12.5 = **158.**

It will be seen that the template now has two inside diameters; a starting point and a finishing point corresponding with the positions on **fig 9-3c.**

In the case of flight **A** these will be points **0 and 1.**

For flight **B** these will be points **1 and 2.**

For flight **C** these will be points **2 and 3.**

For flight **D** these will be points **3 and 4.**

Fig 9-5c shows how the transition is made between the two points.

The difference between the two diameters is divided into a number of equal parts and the **flight circumference** is divided into the same number of equal parts. Radius lines are drawn to intersect with the radial lines of the circumference divisions to plot the transition line. In this case **6** divisions have been used.

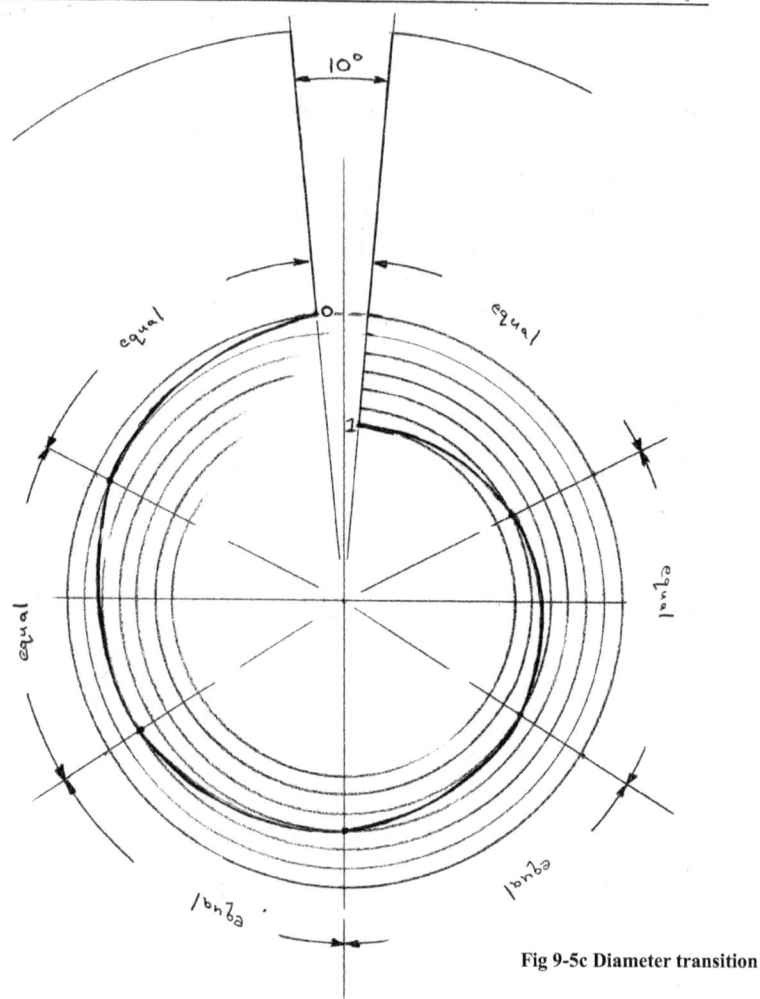

Fig 9-5c Diameter transition

Template B.

Flight height = **120.5.**

Outer circumference = **1648.58.**

Inner circumference = $\sqrt{(275 \times \pi)^2 + 300^2}$ = **914.54.**

Pitch = **300.**

$16.4858 - 9.1454 = \textbf{7.3404}$............ $\dfrac{7.3404}{120.5}$ .... angle = **3.4859°.**

True id = $\dfrac{9.1454}{\tan 3.4859}$ = **150.13.** ..... x2 = **300.26** ...... xπ = circ........... = **943.29.**

True od = 150.13 + 120.5 = **270.63....** x2 = **541.26** ...... xπ = circ........... = **1700.418.**

$\dfrac{914.54}{943.29}$ = **0.9695**.................... $\dfrac{1648.58}{1700.418}$ = **0.9695**........ x 360 = **349.02.**

Cut-out angle = 360 - 349 = **11°.**

Flight height at 1 = **120.5.**

Flight height at 2 = **133.**

Template C.

Flight height = **133.**

Outer circumference = **1648.58.**

Inner circumference = $\sqrt{(250 \times \pi)^2 + 300^2}$ = **840.74.**

Pitch = **300.**

$16.4858 - 8.4074 = \textbf{8.0784}$................. $\dfrac{8.0743}{133}$ .... angle = **3.4758°.**

True id = $\dfrac{8.4074}{\tan 3.4758}$ = **138.419**..... x2 = **276.838** ...... xπ = circ..... = **869.71.**

True od = 138.419 + 133 = **271.419**..... x2 = **542.838**.... xπ = circ..... = **1705.375.**

$\dfrac{840.74}{869.71}$ = **0.96669**................. $\dfrac{1648.58}{1705.375}$ = **0.96669**........... x 360 = **348.**

Cut-out angle = **360 - 348 = 12°.**

Flight height at **2 = 133.**

Flight height at **3 = 145.5.**

Template D.

Flight height = **145.5.**

Outer circumference = **1648.58.**

Inner circumference = √(225 x π)² + 300² = **767.88.**

Pitch = **300.**

16.4858 - 7.6788 = **8.807**........... $\dfrac{8.807}{145.5}$ .... angle = **3.4638°.**

True id = $\dfrac{7.6788}{\tan 3.4638}$ = **126.86**........ x2 = **253.72** .... xπ = circ..... = **797.1.**

True od = 126.86 + 145.5 = **272.36**..... x2 = **544.72** .... xπ = circ..... = **1711.28.**

$\dfrac{767.88}{797.1}$ = **0.9633** .... $\dfrac{1648.58}{1711.28}$ = **0.9633** .... x 360 = **346.8.**

Cut-out angle = **360 - 346.8 = 13.2°.**

Flight height at **3 = 145.5.**

Flight height at **4 = 158.**

© The Mathematics of Boilermaking by Jim Draper 2017

All four templates can now be made and the transitions between the inside diameters can be marked on each as shown in **fig 9-5c.**

It can be seen that the flight heights on all templates correspond to that of its neighbour.

Knowing how to construct both regular and irregular screw conveyor flights, together with the ability to make a transition between two diameters, should mean that the boilermaker can construct any screw conveyor of any configuration presented to him. The three pieces of information required for each flight will always be:

Outside diameter, inside diameter and pitch.

www.ingramcontent.com/pod-product-compliance
Lightning Source LLC
Chambersburg PA
CBHW060839220526
45466CB00003B/1161